COMPUTER NUMERICAL CONTROL
A FIRST LOOK PRIMER

William W. Luggen

Delmar Publishers

an International Thomson Publishing company I(T)P®

Albany • Bonn • Boston • Cincinnati • Detroit • London • Madrid
Melbourne • Mexico City • New York • Pacific Grove • Paris • San Francisco
Singapore • Tokyo • Toronto • Washington

NOTICE TO THE READER

Publisher does not warrant or guarantee any of the products described herein or perform any independent analysis in connection with any of the product information contained herein. Publisher does not assume, and expressly disclaims, any obligation to obtain and include information other than that provided to it by the manufacturer.

The reader is expressly warned to consider and adopt all safety precautions that might be indicated by the activities herein and to avoid all potential hazards. By following the instructions contained herein, the reader willingly assumes all risks in connection with such instructions.

The publisher makes no representation or warranties of any kind, including but not limited to, the warranties of fitness for particular purpose or merchantability, nor are any such representations implied with respect to the material set forth herein, and the publisher takes no responsibility with respect to such material. The publisher shall not be liable for any special, consequential, or exemplary damages resulting, in whole or part, from the readers' use of, or reliance upon, this material.

Cover Design: Charles Cummings Advertising/Art Inc.

This textbook is dedicated to the students who will use it and learn from it.

Delmar Staff
Publisher: Robert D. Lynch
Acquisitions Editor: Vernon Anthony
Developmental Editor: Denise Denisoff
Project Editor: Thomas Smith
Production Coordinator: Karen Smith/Deb Knickerbocker
Art and Design Coordinator: Cheri Plasse

COPYRIGHT © 1997
By Delmar Publishers
an International Thomson Publishing Company

The ITP logo is a trademark under license.

Printed in the United States of America

For more information, contact:

Delmar Publishers
3 Columbia Circle, Box 15015
Albany, New York 12212-5015

International Thomson Publishing
   Europe
Berkshire House
168–173 High Holborn
London, WC1V 7AA
England

Thomas Nelson Australia
102 Dodds Street
South Melbourne, 3205
Victoria, Australia

Nelson Canada
1120 Birchmount Road
Scarborough, Ontario
Canada, M1K 5G4

International Thomson Editores
Campos Eliseos 385, Piso 7
Col Polanco
11560 Mexico D F Mexico

International Thomson Publishing
   GmbH
Konigswinterer Strasse 418
53227 Bonn
Germany

International Thomson Publishing
   Asia
221 Henderson Road
#05-10 Henderson Building
Singapore 0315

International Thomson Publishing–
   Japan
Hirakawacho Kyowa Building, 3F
2-2-1 Hirakawacho
Chiyoda-ku, Tokyo 102
Japan

All rights reserved. No part of this work covered by the copyright hereon may be reproduced or used in any form or by any means—graphic, electronic, or mechanical, including photocopying, recording, taping, or information storage and retrieval systems—without the written permission of the publisher.

2 3 4 5 6 7 8 9 10 XXX 02 01 00 99 98 97

Library of Congress Cataloging-in-Publication Data

Luggen, William W., 1947–
   Computer numerical control : a first look primer / William W. Luggen.
      p.   cm.
   Includes index.
   ISBN 0-8273-7245-0 (pbk.)
   1. Machine-tools—Numerical control. I. Title.
TJ1189.L83   1997
621.9'023—dc20

96-2895
CIP

CONTENTS

Preface	 v	
Section 1	**GETTING STARTED WITH CNC AND PREPARING YOURSELF** 1	
	Objectives 1	
	Introduction 2	
	Unit 1 First Exposure and General Safety 3	
	Unit 2 What is N/C and CNC? 13	
	Unit 3 From N/C Yesterday to CNC Today 20	
	Unit 4 Why Study This Material? 32	
	Unit 5 Getting and Keeping the Job 45	
	Unit 6 Customer Focus and Importance 50	
	Unit 7 Continuous Improvement, Teams, and Your Involvement 55	
	Unit 8 Math for CNC 64	
	Unit 9 Practicing Your Math 81	
	Key Points 92	
	Review Questions and Exercises 94	
	Shop Activities 96	
Section 2	**CNC OPERATION AND COMMUNICATION** 97	
	Objectives 97	
	Introduction 97	
	Unit 10 How N/C Data is Stored and Used 99	
	Unit 11 Drive Motors, Motion, and Feedback 104	
	Unit 12 Axis Relationships and Control 109	
	Unit 13 Types of CNC Systems 123	
	Unit 14 N/C Data Input and Storage 130	
	Unit 15 N/C and CNC Formats, Codes, and Downtime 142	
	Key Points 150	
	Review Questions and Exercises 152	
	Shop Activities 153	
Section 3	**BASIC CNC FUNCTIONS AND FEATURES** 155	
	Objectives 155	
	Introduction 155	
	Unit 16 Functions Controlled by N/C and CNC 157	
	Unit 17 Sequence Numbers 161	
	Unit 18 X and Y Words 164	

iii

	Unit 19 Feed Rates and Spindle Speeds	168
	Unit 20 Programmable Z Depth	172
	Unit 21 Miscellaneous Functions	178
	Unit 22 Preparatory Functions and Operations Performed	181
	Key Points	188
	Review Questions and Exercises	189
	Shop Activities	191
Section 4	**PUTTING IT ALL TOGETHER—SIMPLE PART PROGRAMMING**	193
	Methods and Examples	193
	Objectives	193
	Introduction	193
	Unit 23 Setting Up the Job and Getting Ready to Run	195
	Unit 24 Simple CNC Lathe Programming	208
	Unit 25 Simple CNC Machining Center Programming	222
	Unit 26 Some Additional Programming Functions	233
	Key Points	247
	Review Questions and Exercises	250
	Shop Activities	252
Section 5	**TOOLING AND FIXTURING FOR CNC MACHINES**	253
	Objectives	253
	Introduction	253
	Unit 27 Cutting Tools and Application Practices	255
	Unit 28 Using Carbide Inserts and Toolholders	270
	Unit 29 Fixture Usage and Basic Principles	282
	Unit 30 Clamping, Workholding, and Setup Tooling	292
	Unit 31 Some Sound Tooling Practices	299
	Key Points	305
	Review Questions and Exercises	307
	Shop Activities	308
Section 6	**ADVANCED CNC CAPABILITIES**	309
	Objectives	309
	Introduction	309
	Unit 32 New CNC Features and Advanced Applications	311
	Unit 33 Language- and Graphics-Based Computer Programming	320
	Unit 34 CAD/CAM Basics and Beyond	332
	Unit 35 Flexible Manufacturing Cells and Systems	343
	Unit 36 Automation Concepts and Capabilities	353
	Key Points	372
	Review Questions and Exercises	375
	Shop Activities	375
Appendix A		377
Appendix B		381
Glossary		387
Index		415

PREFACE

The subject of computer numerical control (CNC) can be both interesting and intimidating, depending on how it is approached. Many CNC texts on the market today try to cover too much material too fast. As a result, students can become frustrated and quickly lose interest in the subject matter and the text.

This text is a "first look" at computer numerical control. It is structured and designed to be a primer on the subject and to provide the beginning student of computer numerical control with a good, solid understanding of the basics, fundamentals, and principles of a highly skill-oriented technology by breaking the subject down into numerous single-topic learning units. It is not designed and written to cover everything about the subject in one text, and it is not the only text you will ever need covering computer numerical control.

My goal for this text was to take a complex subject, break it down into very small, easy-to-read and easy-to-understand learning units, and write it directly to you, the student, in a simplified, "first look" approach. The material is presented in a competency-based manner with continuous reinforcement of learned principles through a simplified approach that takes a complex subject and reduces it to its lowest common denominator of "bite-sized" learning units. This approach makes the text suitable for classroom use as well as individualized instruction.

This text is designed to be a secondary or postsecondary education student's first exposure to computer numerical control. Its focus is to be an introductory approach to the world of CNC beginning with safety, jobs, job titles and responsibilities, opportunities, criticality of the work ethic, customer focus, teamwork, and continuous improvement and then advancing to the technical aspects of the field.

As you page through this text you will notice that the units are small, the print size is larger than normal, there are lots of photos and illustrations, there are "Tech Talk" windows of information, and other features, and the material is written in a very helpful, encouraging manner. The writing style is succinct, with lists, bulleted items, exercises, and "things to do and look for" so the student does not have to read a voluminous amount of material and sort out the "need-to-know" from the "nice-to-know."

The text material is organized into six major sections with 36 individual, "bite-sized" learning units per section. Each section begins with objectives and an introduction discussing succeeding topic relevancy and ends with key points, review questions and exercises for the entire section, and selected shop activities. Individual units begin with an inset of key terms to be discussed in that particular unit. Key terms are italicized and boldfaced the first time they are used in the text. Each unit ends with "Recalling the Facts" questions for that particular unit.

Section review questions are both objective and subjective but designed to be intuitive, competency-based, and promote critical thinking. Some of the exercises are math- and English-related consisting of CNC math-related problems and exercises and answering questions with full sentences using newly learned terms and definitions. Other exercises consist of finding new words and writing their meaning, and writing paragraphs about new topics and information learned. Shop activities are designed to get the student out of the classroom and on to the shop floor to identify types of CNC equipment, safety considerations, tooling, control features, characteristics, and push buttons along with beginning to enter commands into the control and execute some basic commands and functions (with the instructor's supervision).

Key features of the text include:

- uniquely and concisely organized subject matter with six major sections and 36 small, "bite-sized" learning units
- easy-to-read text utilizing large print
- text written directly "to" the student in a mentoring, nurturing, and helpful style
- section introductions and learning objectives
- key terms listed at the beginning of each unit; italicized and boldfaced the first time they are used in the text
- "Tech Talk" inset windows containing highlights of interesting and useful CNC-related information
- "Recalling the Facts" questions at the end of each unit to promote immediate reinforcement of learned terms and concepts
- end-of-section "key points" summary, review questions and exercises, and shop activities
- CNC math units and related exercises
- CNC English-related exercises requiring full sentences and written paragraphs using learned terms and concepts
- over 200 photographs and illustrations
- illustrated glossary

ACKNOWLEDGMENTS

I would like to thank the following reviewers.

Don Alexander
Wytheville Community College
Wytheville, VA

Hans Boettcher
Southwestern Oregon Community College
Coos Bay, OR

Tom Clemans
Sno-Isle Skills Center
Everett, WA

John Connolly
Thurston High School
Springfield, OR

Robert Dixon
Chemeketa Community College
Salem, OR

Robert Gianelli
Regional Occupational Skill Center
Erie, PA

Lynn Lockwood
Siuslaw High School
Florence, OR

William L. Shreve
Fox Valley Community College
Oshkosh, WI

David Sizemore
Monroe County Technical Center
 and Robert C. Byrd Institute for
 Advanced Manufacturing
Sinks Grove, WV

Ronald Smeltzer, CmfgE
Los Angeles Pierce College
Woodland Hills, CA

Bruce Whipple
Trident Technical College
Charleston, SC

Donald G. Yost
Dundage Community College
Baltimore, MD

As our manufacturing world continues to become more complex and competitive, interested students must find manufacturing's high-tech material interesting and, at the same time, easy to learn. My sincere hope is that you will find this text interesting, informative, easy to read and understand, and it will provide you with that beginning spark and early information to initiate further study and learning in the field of computer numerical control.

William W. Luggen

SECTION 1

GETTING STARTED WITH CNC AND PREPARING YOURSELF

OBJECTIVES

After studying this section, you will be able to:

- Identify an N/C or CNC machine tool.
- List some general safety rules and guidelines.
- Describe what N/C and CNC is, what it does, and why it was and is important to manufacturing.
- Discuss how N/C and CNC relates to *you* and your future as you study this material.
- List some key points involved in getting and keeping the job you want.
- Identify two types of customers and explain a customer's value to the success of any business.
- Describe continuous improvement and the importance of employee teams and involvement today.
- Review math for CNC preparation.

What is N/C and CNC?

INTRODUCTION

This is a preparation section. It provides the foundation upon which the remaining sections will build for your successful career in CNC manufacturing.

Up to this point in your study of machine tools, you have mostly concentrated on manual machine tools, that is, those with levers and hand wheels. For the most part, many of these machines are still in use throughout industry and will continue to be used. However, a new method of controlling the machine tool that began back in the late 1950s has rapidly gained in popularity. These machines are called *numerical control* (*N/C*) and *computer numerical control* (*CNC*) machines. They have electronic controls, push-buttons, and small readout displays on them, like a television screen. These electronic controls have, for the most part, replaced levers and hand wheels.

You have probably already used electronic controls and performed some simple programs around your home and in your car by using and programming a VCR and stereo. Using the electronic controls on machine tools is somewhat the same; it's just a question of knowing how they work and what you have to do to make them work.

In this section you will learn what kinds of machines are controlled by N/C and CNC, how to safely work around these and other machines, and the difference between N/C and CNC and how that has changed over time. You will learn what this material could mean to you and your future employment using modern machine tools and what kinds of characteristics an employer is looking for. You will also learn about the importance and value of customers to the success of any business, as well as the value and importance of continuous improvement, teams, and active participation in employee involvement programs.

Lastly, you will learn or review some math in preparation for your CNC learning. Good preparation and a solid background and understanding of math is very important to your success in this interesting and exciting field. Math is very important because you need to know how to program (with numbers) a part in order to make it, how to check whether your program will produce a good part, and how to debug and solve programming problems on the shop floor. So, let's get started!

UNIT 1

First Exposure and General Safety

KEY TERMS

program   *programmer*   *operator*
N/C   *CNC*

FUNDAMENTALS OF N/C AND CNC

You have probably seen, read about, or heard about numerically controlled (*N/C*) or computer numerically controlled (*CNC*) machine tools. They are very easy to identify because they have no hand wheels or levers as do manual machines. N/C and CNC machines have electronic controls attached to them with a built-in readout screen similar to a small TV screen. The screen displays various positions that the machine has and will move to as well as specific codes, commands, and other functions that cause the machine tool to function. Some examples of these machines and their control units can be seen in Figures 1–1, 1–2, 1–3, and 1–4.

N/C and CNC machines are a lot like people; alike in some ways and different in others. N/C and CNC machines are alike in the sense that they both have certain physical characteristics in common (N/C and CNC controls, no hand wheels or levers, ability to be programmed, etc.) just like people share certain common physical characteristics (heads, eyes, arms and legs, etc.). N/C and CNC machines also understand the language that is being communicated to them (a program with instructions for the machine to perform some functions). People understand a language, too, and can act on information to perform tasks and functions.

N/C and CNC machines are dissimilar in that there are many different types of machines with many varied types of N/C and CNC controls and related options that make each machine and control combination somewhat different. People have similar differences because each person is an individual with unique features, personal-

FIGURE 1-1
A modern CNC turning center. (Courtesy of Okuma America Corporation.)

FIGURE 1-2
A CNC machining center. (Courtesy of Cincinnati Milacron, Inc.)

FIGURE 1-3
An advanced CNC 3 spindle profiler. (Courtesy of Cincinnati Milacron, Inc.)

ity, and character. This makes each individual different from any other individual. This is what makes you, you!

In order for you to begin to learn and develop your knowledge and skills in this exciting area, it is very important to learn the basics of N/C and CNC technology. These basics are put together in a "building block" approach throughout this text. That is, one topic or unit that you learn will help you learn the next topic. All of this material is important for your general understanding, your safety, and perhaps your potential employment in this field.

FIGURE 1–4
A modern CNC unit. (Courtesy of Cincinnati Milacron, Inc.)

SAFETY

The first and most important thing to learn about N/C and CNC machines is *safety*. In your previous studies of manual machines you learned a great deal about safety from your instructors and from your previous textbook. All of these rules, guidelines, and procedures, for the most part, still apply. You must remember that with manual machines, nothing happened until you turned the spindle on, moved the cutting tool into the workpiece, and began to cut metal. All of this was completely under your control. *You* started and stopped the spindle. *You* decided what tool to use first. *You* decided where to begin to cut, and how much of a cut to make. All of these little decisions were under *your* control.

With N/C and CNC machines, all of these moves may or may not be totally under your control. The machine moves are typically thought out and assembled into a "package" of moves called a ***program,*** which is in place *before* the machine begins to cut metal. All the moves that make up the program are input to the electronic control by an ***operator*** or a ***programmer***. This means that the machine is being directed to perform its cutting functions either by yourself or someone else who made the program and hopefully knew what they were doing. If you or that someone else make a serious mistake, a machine crash can occur causing serious damage to the machine, the part, the tool and tool holder, and possible injury to yourself or someone else. Therefore, N/C and CNC machines require a new and higher level of concern and awareness as far as safety is concerned.

This is not an attempt to scare you about this new equipment, but you must begin your understanding of the differences between manual and N/C and CNC machine tools with safety. In many ways it's just like learning to drive a car. The first things you learn are how to make it go, how to make it stop, how to steer it, and the rules of the road. With N/C and CNC machines you must also know how to make them go, how to make them stop, how to steer them (give them instructions through the program), and the "rules of the road."

We will discuss programs and input to those controls in more detail later, but first let's talk more about safety and your responsibilities. Whether you are programming or operating the machine, *you* are still responsible for its actions. If you are programming the machine either at the machine or off-line in an office, you must make sure that you are always aware of where the cutting tool is at all times and that all moves are programmed safely. We will discuss programming in more detail later in the text, but for now it's important for you to know that safety with N/C and CNC machines is not just the responsibility of the person running the machine.

If you are operating the N/C or CNC machine, you are responsible for *all* of its actions, whether you programmed the machine or not! You are responsible for your own safety and for the safety of those around you. Therefore, you must always be alert and aware of the machine, the condition of the machine, how the material is being cut, the cutting conditions, and the tool and workholding conditions and situation. As you gain knowledge about N/C and CNC machines, you will learn that these machines move at very high feeds and speeds and are very unforgiving; they will do exactly what you tell them to do. If a crash happens because of a particular move you or someone else programmed, the machine does not know that; it will simply try to do what you tell it. Therefore, be alert, anticipate dangerous moves and conditions, and pay attention at all times around *any* machinery.

Now let's list and discuss some safety rules for N/C and CNC machines, as well as some general shop safety.

1. You probably have been assigned safety glasses from your previous metalworking classes and are aware of what can happen when you don't wear them. Therefore, wear safety glasses and preferably safety (steel toe) shoes at all times when working in a shop environment.

2. It's just as important *not* to wear certain articles as it is to wear others. The list of items not to wear around either manual or N/C machinery includes, but is not limited to: neckties, long sleeves, wristwatches, rings, gloves, identification bracelets, and the like. In addition, shirttails should always be tucked in.

3. Always maintain good housekeeping. In other words, keep your workplace clean, neat, and orderly. Make sure the area around automated equipment is well lighted, dry, and free

FIRST EXPOSURE AND GENERAL SAFETY • 9

from clutter and obstructions that you or someone else could fall over.

4. Pay attention to all warning lights, beacons, and audible signals. If a signal, sound, or warning light alerts you, know what to do or contact your supervisor or instructor.

5. Hand tools, clamps, gauges, measuring tools, etc. should be kept off all operating equipment (whether manual or N/C) and all their moving units.

6. Avoid bumping any processing equipment or control units.

7. Never perform grinding operations near any piece of automated equipment and all its moving units. Abrasive dust gets in and clings to moving parts and will cause undue wear, inaccuracies, and possible failure of affected parts.

8. Never place hands near a revolving spindle. Keep hands away from moving equipment units.

9. Always perform setup, tool, and fixture adjustments, as well as workpiece checks, with the spindle stopped.

10. Always double-check and make sure parts and fixtures are securely clamped before beginning machining.

11. Always double-check all tools before using them and use only properly sharpened tools.

12. Do not use compressed air to blow chips from the part or to clean equipment surfaces, cabinets, controls, or the general work area.

13. When handling or lifting parts or tooling, follow your company or school guidelines for correct procedures. Remember to lift with your legs, not your back. Get help if you think something is too heavy or use a crane and follow proper procedures.

14. Work platforms around machinery should be sturdy and must have antislip surfaces.

15. When handling tools or changing tools by hand, use a glove or shop rag. Avoid contact with cutting edges. Do not operate equipment while still wearing gloves.

FIGURE 1–5
The "master stop" button is clearly displayed for quick and easy access if an emergency occurs. (Courtesy of Mazatrol T Plus CNC Control.)

16. Always exercise caution when changing tools and avoid interference with the fixture or workpiece.

17. Never manually operate any automated equipment without proper training or supervision. Consult the specific operator's manual for that particular machine and control type.

18. Always know exactly where the "master stop" is located on any machine tool before operating it. On N/C and CNC ma-

chines it is usually the big red button on the control. One can be seen in Figure 1–5. If you push this button in an emergency condition, the machine will stop all action and shut down hydraulics and electronics.

19. Never attempt to program any automated equipment without proper training or supervision. Consult the specific programming manual for that specific machine and control type.

20. Never attempt to manually remove a tool from the spindle of an N/C or CNC machine tool. Never attempt to remove a tool from the tool storage matrix of an N/C or CNC machine tool unless the N/C or CNC control and/or machine is switched to the manual mode.

21. Never lose alertness, concentration, or doze around operating machinery. Machines that operate automatically have a tendency during repetitive production runs to lull you into a state of drowsiness. Remember, operating N/C and CNC machinery is just like driving a car; you must pay close attention to what you are doing and what the machine is doing at all times!

22. Electrical compartment doors should be opened only for electrical and/or maintenance work. They should be opened only by experienced electricians and/or qualified service personnel.

23. Safety guards, covers, and other devices have been provided for your protection and that of your fellow students and co-workers. Do not operate any equipment with these devices disconnected, removed, or out of position. Operate equipment only when they are in proper operating condition and position.

RECALLING THE FACTS

1. N/C and CNC machines are easy to identify because they have no _____ like manual machines.

2. The first and most important thing to learn about N/C and CNC is _____.

3. Machine moves are typically thought out and assembled into a "package" called a _____.

4. Learning N/C and CNC machine tools is a lot like learning to drive a car. The first things you learn are how to make them _____, how to make them _____, how to _____ them, and the "rules of the road."

5. When programming an N/C or CNC machine, you must make sure that you always know where the _____ is at all times and that all moves are programmed _____.

UNIT 2

What Is N/C and CNC?

KEY TERMS

words   *MCU*   *programming*
block   *instructions*

INTRODUCTION TO N/C

The two words that have had the biggest impact in manufacturing since the 1960s are *numerical control*. Numerical control has revolutionized manufacturing, lowered costs, improved quality, shortened lead times, and dramatically improved productivity.

Numerical control is basically control of machine tools by numbers. These numbers have certain letters that go with the numbers. Together the numbers and letters, actually called *words* in N/C language, make up the codes, commands, position values, and functions that direct the activity of the N/C machine tool. The words are entered into the electronic control by various means, which we will discuss later, and are translated into pulses of electric current and other output signals that activate motors to run the machine. These words initiate action from positioning the machine table and beginning a drilling operation to changing a tool and turning the spindle and coolant on, as well as many other functions.

The codes, commands, positions and functions (words) are essentially individual *instructions*, sometimes called *blocks*. One block of information is one instruction. Each block or instruction may contain one or more words, as seen in Figure 2–1. They are assembled into a "package" called a *program*, as we discussed in Unit 1. A very simple program is shown in Figure 2–2. Don't worry about what the program will do or what the words mean at this time. Just look at the overall program and the individual words that make up the program. The program is an organized sequence of events that

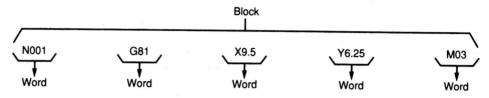

FIGURE 2-1
A CNC block or instruction and individual words.

```
            N10   T1     M6
            N20   G0 X2. Y1.
            N30   G81 R.1 Z-.5 F3.
            N40   X4.
            N50   X6.
            N60   Y3.
            N70   X4.
FIGURE 2-2  N80   G80
A simple CNC program.  N90   G0 Z5. M30
```

directs the activity of the machine tool. It represents the thoughts, ideas, and actions of the programmer who wrote the words in the right order to make up the program. Programs may be stored by various means, which we will talk about later, and reused over and over again to make a particular part or a batch of identical parts.

PROGRAMMING

The concept of *programming* is the essence of what N/C and CNC is all about. Before any words are written or coded into a machine's *MCU (machine control unit)*, a programmer must have a good idea of what he/she wants to do and some prior manufacturing skill and experience to draw from.

This is very important and is why you learn manual machines before you learn N/C; you must understand the metalcutting process, feeds, speeds, cutting tools, machines, machining methods, and so forth in order to make good decisions about how to instruct and direct the N/C or CNC machine. Learning the coding

> **"Tech Talk"**
> The total number of N/C and CNC machines in the U.S. machine tool population is still relatively small. Recent estimates are that out of roughly a three million machine tool universe, the N/C and CNC units number only about 200,000, or slightly over 6.5%.

and the individual words is easy. Learning how to put them all together to make a good program that cuts an accurate part takes skill, knowledge, training, and experience.

In Unit 1 we discussed similarities and differences in N/C and CNC machines. These similarities and differences also apply to programming. Although programming standards exist, programming for all N/C and CNC machines is not the same. There are different types and styles of machining centers (vertical and horizontal), turning centers (vertical and horizontal), milling machines, boring machines, special machines, etc. from many different manufacturers. Some of these can be seen in Figures 2–3, 2–4, and 2–5. There are also different types and styles of control units from different man-

FIGURE 2–3
A vertical (vertical spindle) turning center. (Courtesy of Giddings and Lewis, Incorporated.)

FIGURE 2–4
A horizontal (horizontal spindle) machining center. (Courtesy of Mazak Corp.)

ufacturers. Mixing and matching these different machines and controls together yields a staggering array of combinations available, each with its own "language," so to speak. Whoever is doing the programming must be familiar with the specific machine and control combination or language for which he or she is programming. It's just like trying to communicate with someone from another country in a different language. If you don't understand the language well enough to communicate what you want or what you want the person to do, that person can't act on what you want. The

FIGURE 2–5
A vertical machining center; Mazak VTC-20B Vertical Machining Center. (Courtesy of Mazak Corp.)

important thing to remember here is that among different N/C and CNC machine and control types, all are not programmed exactly the same; there are similarities and there are differences.

If programming is the essence of what N/C and CNC is all about, then visualization is the essence of what programming is all about. Programmers must be able to visualize in their "mind's eye" what the machine will do as they choose what words to use and how to arrange them in the program. Without the ability to foresee, understand, and anticipate each individual move the cutter makes in relation to the part, serious crashes, accidents, and part inaccuracies can occur. Therefore, programmers must talk the specific language of the machine control unit (MCU). Being able to program N/C and CNC machine tools well comes only from training, practice, and experience. We will talk much more about programming later in the text.

Now that we have discussed what N/C and CNC is, we also need to talk about what it is not. First, the actual N/C machine tool can do nothing more than it was capable of doing before a control unit was joined to it. There are no new metal-removing techniques or principles involved. N/C and CNC machines position and drive the cutting tools, but the same milling cutters, drills, taps, and other tools still perform the cutting operations. Cutting speeds, feeds, and tooling principles must still be followed. Given this knowledge, what's the real advantage of numerical control? Primarily, the idle time, or the time required to move into position for new cuts, is limited only by the machine's capacity to respond. Because the machine receives commands from the machine control unit, it responds to just what it is told. The actual "time in the cut," or the time that the machine actually cuts metal, is therefore much higher than on a manually operated machine. The second point is that numerically controlled machines can initiate nothing on their own. The machine reacts and responds to commands from the control unit. Without some form of input medium (preprogrammed or at-the-machine data entry), the machine and control unit will do nothing on their own.

RECALLING THE FACTS

1. _____ is basically control of machine tools by _____.
2. The numbers and letters, called _____ in N/C language, make up the codes, commands, position values, and functions that direct the action of the machine tool.
3. Each _____ or instruction contains one or more _____.
4. _____, consisting of codes, commands, positions, and functions are assembled into a "package" called a *program*.

5. The _____ is an organized sequence of events that directs the activity of the machine tool.
6. The concept of _____ is the essence of what N/C and CNC is all about.
7. The machine tool receives commands from the _____ unit, or _____, for short.

UNIT 3

From N/C Yesterday to CNC Today

KEY TERMS

hard-wired
executive program
software-based
microprocessor
load tape

ORIGINS OF N/C

Looking back at the history of numerical control and where and how it all started, many people agree that the first real numerical control (N/C) machines were the 1725 knitting machine and the player piano, introduced around 1863. The knitting machine and the player piano are considered the true forerunners of modern numerical control. Actually, numerical control as we know it today began in 1947. John Parsons of the Parsons Corporation, based in Traverse City, Michigan, began experimenting with the idea of generating thru-axis curve data and using that data to control machine tool motions. Numerical control originated as Parsons discovered a way of coupling computer equipment with a jig borer.

In 1949 there was a demand for increased productivity for the U.S. Air Materiel Command as the parts for its airplanes and missiles became more complex. In addition, the designs were constantly being changed and revised. A contract was granted to the Parsons Corporation to search for a speedy production method. In 1951, the Massachusetts Institute of Technology (MIT) assumed this work. In 1952, MIT successfully demonstrated a primitive model of the N/C machine of today. The machine was a vertical spindle Cincinnati Hydrotel with a very large lab-constructed control unit. The machine successfully made parts with simultaneous thru-axis cutting tool movements. Although Parsons came up with the idea of numerical control, MIT actually gets credit for coming up with the term *numerical control*.

In 1955, at the National Machine Tool Show, commercial models of numerically controlled machines were displayed, ready for customer acceptance. Shown were different contour-milling machines, which were very expensive. Some machines required trained and skilled mathematicians and the primitive computers of the day to produce tapes. By 1957, numerically controlled machines were slowly beginning to be accepted by industry; several had been installed and were in use in production applications.

MODERN CNC MACHINES

Both machine tool and control/electronics technology have evolved rapidly since the introduction of numerical control. Capabilities have increased dramatically while machine tool and control unit size and cost have decreased. Machines that used to cost $100,000, for example, now can be bought for $30,000 or less. Machines and controls have also become safer, easier to use, and in some cases, can be unattended. This has helped manufacturers become more productive, improve their quality, and make more money in their work. New types of drive motors, high-speed spindles, and other types of electronic and mechanical breakthroughs have all combined to boost performance capabilities considerably higher than those achieved with manual machines. In one day it is now possible to take delivery of a CNC machine tool, set it up, and produce parts within tolerance and without problems.

> **"Tech Talk"**
>
> John Parsons, for his work on the initial development of N/C, was fired by his own company. He was only a minority stockholder at the time. He was eventually re-hired and N/C did become successful. Once N/C became successful, Parsons eventually sold his patent rights to the Bendix Corporation and used the money to buy control of his own company.

At present, there are many types of N/C and CNC machines producing parts in manufacturing plants. They range in age from some of the earlier models (Figure 3–1) to modern and advanced CNCs (Figures 3–2, 3–3, and 3–4). Not all N/C and CNC machines are of the metal-cutting variety. Some are hole punching machines, tube benders, inspection machines, welding machines, flame cutters, co-

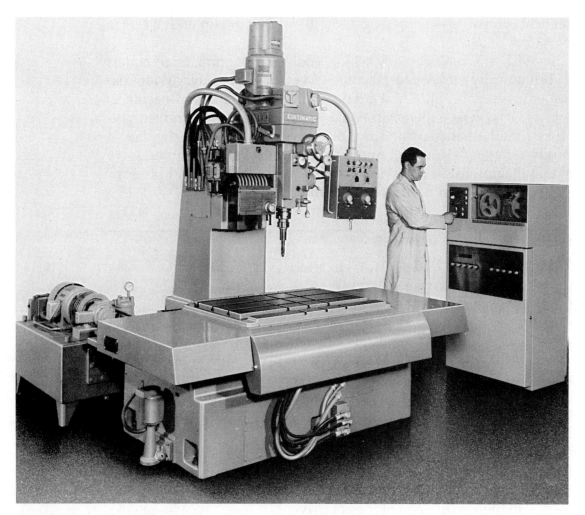

FIGURE 3–1
An older N/C machine tool. (Courtesy of Cincinnati Milacron, Inc.)

FIGURE 3–2
A horizontal CNC turning center. (Courtesy of Okuma America Corporation.)

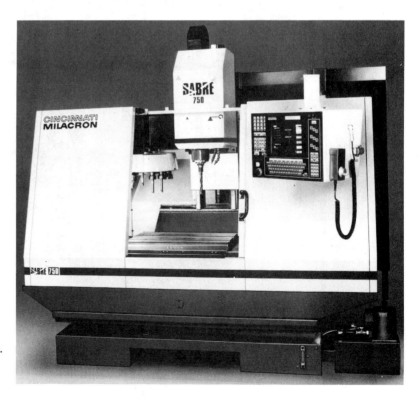

FIGURE 3–3
A vertical machining center. (Courtesy of Cincinnati Milacron, Inc.)

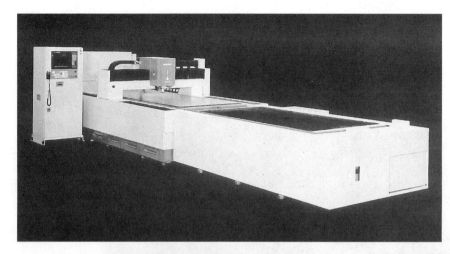

FIGURE 3–4
A CNC laser machine tool. (Courtesy of Mitsubishi Laser.)

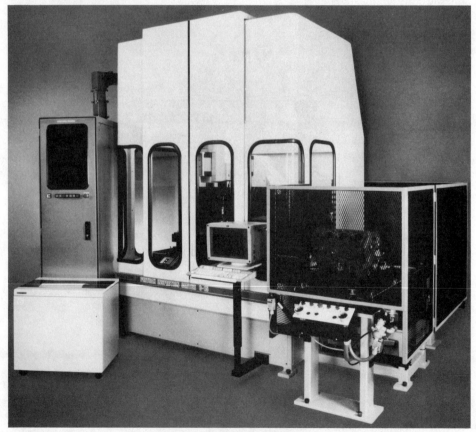

FIGURE 3–5
A CNC coordinate measuring machine (CMM). (Courtesy of Giddings and Lewis, Incorporated.)

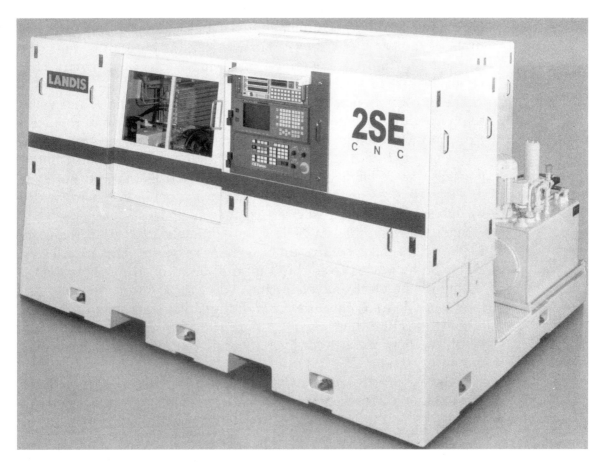

FIGURE 3–6
A CNC grinder. (Courtesy of Landis Div. of Western Atlas Inc., Waynesboro, PA.)

ordinate measuring machines, and precision centerless and center-type grinders, as seen in Figures 3–5 and 3–6.

ELECTRONIC ADVANCEMENTS

Although machine tools themselves have advanced considerably since Parsons and MIT first began their work, no greater change has occurred in the field of numerical control than with the machine control unit (MCU). Controls have progressed from the bulky tube types (just like TVs used to have tubes) to the microprocessor-based units of today. A *microprocessor* is a central processing unit (the computer's "brains") contained on a single integrated circuit chip about the size of your fingernail, or smaller.

The introduction of the transistor, solid state circuitry, and eventually modular or integrated circuits (ICs) led the way in this electronic revolution and enabled machine tool manufacturers to combine many individual circuits into a single microchip and to offer features and reliability to N/C machine controls that did not exist in the early versions. As a result, control capability and reliability increased as size and cost decreased.

Since the beginning of N/C, approximately 1954, until the early 1970s, all machine control units were *hard-wired*. This meant that all of the logic was fixed, built-in and determined by the physical wiring of the electronic elements within the control unit. Because they were "wired" into the control, these elements then controlled all functions that the control could do, such as which tape format it would accept, absolute or incremental positioning, and character code recognition.

In the early 1970s, more capable and less expensive electronics began to emerge. These types of computer elements, complete microprocessors (a computer on a chip), became part of the control units. Functions that were solely the result of hardware design and "wired" into the control became resident in computer logic on the microprocessor within the control unit. Thus, the evolution of numerical control was about to make a major change and advancement from the hard-wired N/C units to advanced CNC (computer numerical control) units.

SOFTWARE-BASED CONTROLS

Because the new CNC controls contained microprocessors allowing their internal logic to be changed, they were called *software-based* controls. Essentially, the internal control logic could be changed or additional logic could be added to make a machine control unit recognize different tape formats, or expanded to add new codes or functions.

Software-based CNC controls were first displayed at the 1976 Machine Tool Show. Soon after, the hard-wired controls began to be replaced with computer logic controls which had more capability, were no more costly, and could be programmed for a variety of functions at any time. With software-based logic, all the codes, functions, and commands were not locked in at time of manufacture. It was simply a matter of how the computer software logic within the control unit was programmed to read the various codes, functions, commands, etc.

The physical components of the software-based CNC units are the same regardless of machine type. It is not the control unit elements, but rather the *executive program*, or *load tape*, that gives the control its "brains" and makes it "think" like a machining center or lathe. The executive program is logic-based and permanently resides within the CNC; it can be revised, updated, or modified at any time. New codes, commands, and functions can be introduced by simply reading or programming them into the executive portion of the control unit computer. Changing functions in a hard-wired control would involve physically changing the particular electronic controlling elements within the control unit wiring. The executive program is supplied by the control unit manufacturers. In most cases, the user does not attempt to alter the executive program in any way. A general comparison of the characteristics of N/C versus CNC is given in Figure 3-7.

The advent of computer numerical control in the mid-1970s has resulted in a much greater acceptance of numerical control in industry. This is because, unlike the N/C versions of the 1950s and 1960s, CNC today does not require significant training. CNC machines have become easier to use with menu-selectable displays, advanced graphics, and conversational English programming (see Figures 3-8, 3-9, and 3-10). What used to be reserved for master-level machinists can now be rapidly learned and understood by entry-level machinists. Virtually all numerically controlled machine tools manufactured today are of the CNC type.

N/C	CNC
Hard-wired	Software logic based
Control logic fixed (codes, functions, commands, etc.)	Control logic based in resident microprocessor
Internal control logic not changeable except through circuit board changes	Executive program changeable; makes control "think" like a machining center or turning center
No memory	Memory capacity with computational and compensational ability
Must be externally programmed and tape punched; no program created at machine	Program entered via direct line from external computer, floppy disk, cassette, punched tape, or manual data input
Programs stored on punched tape or cards	Programs stored on external computer and downloaded on demand via direct line; also stored in memory or on floppy disk or cassette
Tape must be recycled for every part	Memory capacity holds program; program executed from memory

FIGURE 3–7 A comparison of N/C versus CNC general characteristics.

FIGURE 3–8 A CNC display screen. (Courtesy of GE Fanuc Automation.)

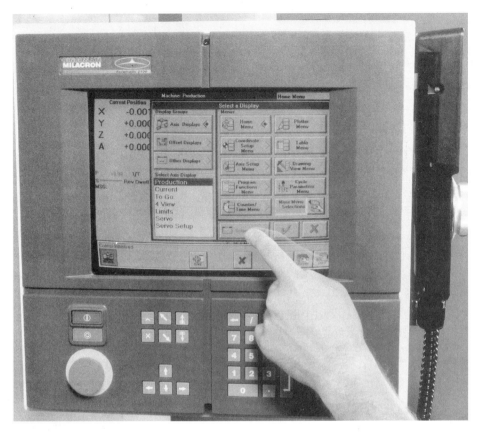

FIGURE 3–9
A CNC's display screen showing its menu–selectable format. (Courtesy of Cincinnati Milacron, Inc.)

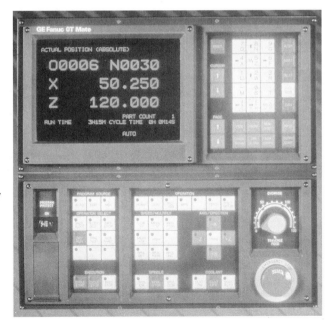

FIGURE 3–10
CNCs display screens show graphics, functions, commands, positions the machine is currently at, positions the machine will be moving to, error messages, and other information. (Courtesy of GE Fanuc Automation.)

RECALLING THE FACTS

1. Numerical control as we know it today began around _____ by _____ but _____ actually gets credit for coming up with the term *numerical control*.
2. Control units have progressed from the bulky tube types of yesterday to the _____-based units of today.
3. _____ means that all the decision logic was built in and determined by the physical wiring of the electronic elements.
4. _____ CNC controls allow the CNC internal decision logic to be easily changed or modified.
5. The executive program, or _____, is supplied by the control unit manufacturer.
6. As electronics evolved since the beginning of N/C, control capability and reliability _____ as size and cost _____.

UNIT 4

Why Study This Material?

KEY TERMS

operator
setup person
accuracy
operator/programmer
programmer
repeatability
tooling engineer
methods engineer
reliability

THE STUDENT OF CNC

Now that you know what N/C and CNC is, and before we get into more detail about this material, let's talk about why it's important to study this material in the first place.

The importance of studying CNC will be put in terms of what it means directly to you and also to users of the technology. In terms of what it means directly to you, it can best be summed up by a quote from an unidentified source: *"Man is user of tools; those who recognize the tools of tomorrow and learn how to use them today, assure themselves of a place in tomorrow's prosperity."* Write these words on a piece of paper and tape the paper somewhere where you must see it every day.

For you, studying and learning CNC, the "tools of today and tomorrow" can:

- help you get a job or get a better job in a manufacturing company that uses N/C and CNC
- improve and upgrade your skills
- increase your pay
- improve you and your family's standard of living
- increase your self-confidence, self-esteem, and self-worth
- increase your value to yourself and to your employer
- provide you with an in-demand skill that can be transferred

to another company should you lose your job at some time in your life
- develop an interest that drives you to get more training and leads to more in-depth knowledge and experience (and therefore opens up opportunities leading to increased pay)

People with CNC experience have been and will continue to be in demand. This is because CNC machines are the relatively new "tools" of manufacturing. Anyone who understands them and knows how to program and operate them typically has a good chance of being employed. The reasoning here is that these machines require a higher level of thinking to use than do manual machines. They are also more complex because they require a program containing all the machine and cutter movements to be completed and "thought out" ahead of time, before the machine actually begins to cut metal.

CAREER OPPORTUNITIES

By this time you are probably wondering, "If I am going to learn this material, what types of jobs are available? Which jobs pay the most? How much knowledge and experience do I need to get one of these jobs?" The typical N/C and CNC jobs associated with most manufacturing companies are: operator, operator/programmer, setup person, programmer, tooling engineer, and methods engineer. Keep in mind that these are only the N/C related jobs and that specific names, titles, and responsibilities associated with these jobs may change from one company to another. Let's talk a little about each one.

> **"Tech Talk"**
>
> Some vocational/technical schools in the U.S. teaching CNC have a 100% success rate in placing their CNC graduates in jobs.

Operator. A CNC operator (Figure 4–1) is basically a parts loader and unloader. Typically, this is the lowest level of skill in the N/C and CNC position chain but, nevertheless, a very important position. The operator must be able to read, understand, and interpret the program manuscript, and know how to interface with the machine and the MCU. In addition, he or she must be able to locate the

FIGURE 4-1
An N/C or CNC operator basically loads and unloads parts. Their position is important because they are the final human link in the N/C chain that determines whether the part coming off the N/C machine is good or bad. (Courtesy of Giddings and Lewis, Inc.)

parts properly in the holding fixture time after time, understand machining and tooling fundamentals, and be able to detect and anticipate a host of problems that could happen in the machining of the part. In some companies, the operator is also responsible for the part setup in addition to loading and unloading. Programming is typically done off-line (away from the shop and in the office) by part programmers. The operator is the last human link in the N/C process that determines whether the part comes off the machine good or bad.

Operator/programmer. The operator/programmer does basically the same work as the operator, with the addition of doing at-the-machine programming (Figure 4–2). In addition to understanding the machining, tooling, etc. that the operator has to know, the op-

FIGURE 4–2
The operator/programmer loads, unloads, and sets up his or her own machine in addition to programming at the machine. (Courtesy of Cincinnati Milacron, Inc.)

erator/programmer must also be skilled at print reading and at programming his or her particular machine and MCU. Typically, this individual has gone to school at the equipment manufacturer's plant to learn how to program and operate the machine. The operator/programmer generally works in a smaller job shop where the added expense of having off-line programmers cannot be justified by management and the part complexity and the volume of similar parts to be machined are relatively low. Consequently, the operator/programmer programs his or her own work and then runs the job, often storing the completed output in case the job has to be run again at a later date. This is a higher level position because the operator/programmer takes on the added responsibility of programming the machine in addition to running the machine.

Setup person. The setup person (Figure 4–3) is responsible for setting up the CNC machine tool. This generally includes properly loading the fixture, if any, loading the tools and making sure they are all correct, making sure the program is correct and calling it up or loading it into the control, loading and running the first part, getting it inspected, making any tool adjustments, and then "OK'ing" the job. Once the setup person has "OK'ed" the first part, he or she then turns the job over to the operator who runs production for the rest of the parts. The setup person moves around the shop setting up new parts to be machined and helping operators with various problems. His or her primary responsibily is to keep the machines in production. The setup person is a very knowledgeable individual who understands various N/C equipment, tooling, fixturing, programming, machining, and how to trouble-shoot shop problems. This is also a higher level position because the setup person must be familiar with all types of machines and controls the shop has in addition to knowing how to quickly understand and solve shop problems related to machines, tooling, fixturing, programming, and machining.

Programmer. This is a higher level position in the N/C chain of jobs and job skills. N/C programmers are typically salaried employees whereas operators and operator/programmers are paid by the hour, although this is not always the case. Programmers typically program off-line away from the shop, as seen in Figure 4–4, and are respon-

FIGURE 4–3
The setup person is responsible for setting up the N/C and CNC machines in the shop, running the first part, having the part inspected and approved, solving shop-related problems, and helping the operators. (Courtesy of Cincinnati Milacron, Inc.)

sible for knowing and understanding several different machine and control types. Programming requires a high degree of patience, knowledge, and the ability to concentrate. Programmers must be able to read and interpret part prints and be good at mathematics and geometry in addition to being computer literate, knowledgeable in the computer language or graphics-based system used, and have

FIGURE 4–4
CNC programmers generally program CNC machines off-line in an office but will "prove out" (make sure their program makes a quality part) their N/C programs when the part they are programming is ready to be run on the machine. (Courtesy of International Business Machines Corporation.)

a good understanding of tooling and machining. Most importantly, though, is the ability to "visualize," in three dimensions in their minds, what is happening on the machine while they are programming the part.

Tooling engineer. This is another higher-level position in most manufacturing companies. Tooling engineers are typically responsible for fixture design and seeing that the fixture is manufactured and assembled to specifications. Some tool engineering responsibilities extend to providing for cutting tools, holders, carbide inserts, etc. and include purchasing these tools and components from outside vendors while maintaining their company's own quality standards (Figure 4–5). Tooling engineers must be very good at reading and interpreting part prints, understand dimensioning and tolerancing requirements, have extensive experience with cutting tools and workholding methods and, in addition, know machining and machining concepts.

FIGURE 4–5
Tooling engineers are typically responsible for fixturing and tooling; this includes making sure that the fixture is designed, manufactured, and assembled to specifications and additionally are responsibile for cutting tools, holders, adapters, etc. (Courtesy of International Business Machines Corporation.)

Methods engineer. Sometimes called a *planner* or a *manufacturing engineer*, these salaried individuals are highly experienced and responsible for determining the machining methods, specific processes, and specific machines that will manufacture the various parts. In many cases, the methods engineer, as seen in Figure 4–6, has the added responsibility of part programming and also some tooling. This is dependent on how the manufacturing and/or manufacturing engineering is organized and how the work is structured. This position is probably the highest-level position in the N/C and CNC chain of positions and also the most demanding. Methods engineers need all of the knowledge of the programmer and tooling engineer positions in addition to knowing and understanding a variety of machine tools and manufacturing processes for both manual and N/C machines.

WHY STUDY THIS MATERIAL? • **39**

FIGURE 4–6
Methods engineers, sometimes called planners or manufacturing engineers, are typically responsible for determining the machining methods for all part manufacturing machines and processes; in some cases this includes N/C programming and tooling responsibility. (Courtesy of International Business Machines Corporation.)

Manufacturers must make good use of all their equipment and employees (operators, programmers, tool engineers, methods engineers, and others) in order to make their products and provide a profit. If a manufacturing company buys CNC machines over manual equipment, then it only stands to reason that CNC must be able to do some things better than manual machines.

CNC VS. MANUAL MACHINES

Three things that N/C and CNC machines can do better than manual machines are: increase accuracy, improve repeatability, and provide higher reliability. Let's discuss each of these in more detail.

1. *Accuracy.* CNC machines are more accurate. Accuracy is defined as a measure of the extent to which a part is free from error. A machinist may be able to produce a part that is free from error. Many machinists can work to tolerances of .001 in. on an old, worn machine—if they are very familiar with that machine. A tremendous amount of experience and time is needed to be able to work consistently to this degree of accuracy. N/C and CNC machines can work to much closer tol-

erances than manual machines and the accuracy is built into the machine.

2. ***Repeatability.*** Repeatability is a measure of the differences among the same dimensions of each piece machined. In other words, how close does one piece compare with another? In essence, a machinist on a manual machine may be able to produce good accurate parts time and time again, but how close are the actual dimensions on one part as compared to the others? How much the parts vary from one to another is the operator's or the machine's repeatability. CNC has great repeatability (provided the parts are located properly).

3. ***Reliability.*** Reliability is the measure of dependability, or the extent to which the machine can be consistently relied upon. When a manufacturer buys a CNC machine tool, he wants it to work and perform. If you buy a car, you want it to work and perform, don't you? After all, you spent your money for it just like the manufacturer spent his money for the machine. Just like a car, if a CNC machine tool is "down" and not working, it causes problems and can't be productive. Therefore, reliability of N/C and CNC equipment is extremely important to manufacturing companies.

There are other reasons why CNC machines are better and why manufacturing companies buy them. Some of these reasons are:

- The idle time, or time for the machine to move into position for new cuts, is limited only by how quickly the machine can rapid to position. Because the machine receives commands from the MCU, it responds without hesitation. Therefore, the actual time that the machine is cutting and making chips is much higher.
- N/C makes it possible to produce even the most complex shapes without extremely high costs. With manual machines, complex shapes would require lots of expensive tools and fixtures. Also, costs associated with manual machines increase as part tolerances become tighter.
- With manual machines, it is often too costly to make changes after tooling is prepared. With CNC, expensive tooling is elim-

inated and making many changes only amounts to changing the program.
- Complex workholding devices such as jigs and fixtures are not required in all cases. For most operations, the simplest form of clamping or fixturing device is all that is needed.
- Selection of feeds and speeds, sequence of cutting operations, and so forth, that used to be up to the operator to determine are now under the control of the programmer. This saves time and money because once the job is ready it can be set up on the machine and run.

Additional advantages of CNC are:

- *Reduced scrap and rework.* Errors due to operator fatigue, interruptions, and other factors are less likely to occur on N/C machines.
- *Improved production planning.* CNC machines can often perform, at one setting, work that would normally require several conventional machines.
- *Reduced space requirements.* Since fewer jigs and fixtures are used, the actual storage requirements of these expensive tools are reduced.
- *Simplified inspection.* Once the first piece has passed inspection, minimal inspection is required on subsequent parts.
- *Lower tooling costs.* There is less need for complex jigs and fixtures.
- *Reduced lead time.* This is a result of lower tooling costs.
- *Greater ease in accomplishing complex machining operations.* This is due to advanced machine control and programming capabilities.
- *Increased machine utilization.* This means that the CNC machine runs more and stays "in the cut," resulting in less idle time.
- *Reduced parts inventory.* With CNC, inventory is reduced because small lot sizes can be run quickly with less advance notice and lead time.

LIMITATIONS OF CNC

If CNC was the answer to all manufacturing's problems, there would be no need for manual machine tools. This is not the case. There will always be a need for some manual machine tools because not all parts or part manufacturing operations lend themselves to CNC machines or operations.

Some things you need to know about CNC machines that can be considered limitations to the use of CNC (and can perhaps be called disadvantages) include:

1. CNC machines will not cut metal any faster than manual machine tools. CNC machines only position and drive the cutting tools. Running faster feeds and speeds as well as using new and improved cutting tools can be done on manual machines as well as CNC machines.
2. CNC does not eliminate the need for expensive tools. Some jobs require special and expensive fixtures and cutting tools. Although the cost of CNC machines has come down considerably over the last several years, some are still very expensive and difficult to justify today.
3. CNC will not totally eliminate errors. You can still push the wrong button, make incorrect alignments, or fail to locate parts properly in a fixture. Some of these errors can be reduced by careful and extensive training.
4. CNC should only be considered and used where it will produce the parts better, faster, and more accurately than manual machine tools. This is not always the case.

RECALLING THE FACTS

1. Studying and learning N/C and CNC can:
   - improve and upgrade _____.
   - increase your _____, _____, and self-worth.
   - increase your _____ to yourself and your employer.
   - provide you with an in-demand _____.

2. The _____ is the last human link in the N/C process that determines whether the part comes off the machine good or bad.
3. The _____ loads and unloads parts in addition to doing at-the-machine programming.
4. The _____, typically a salaried person, programs the CNC machine in an off-line office away from the shop.
5. The _____ person moves around the shop setting up new parts to be machined and helping operators with various machining problems.
6. The _____ is responsible for fixture design and seeing that the fixture is manufactured and assembled to specifications.
7. The _____ is sometimes called a *planner* or *manufacturing engineer* and is responsible for determining the machining methods, specific processes, and specific machines that will manufacture the various parts.
8. _____ is defined as a measure of the extent to which a part is free from error.
9. _____ is a measure of the differences among the same dimensions of each part machined.
10. _____ is a measure of the dependability of the extent to which the machine can be consistently relied upon.
11. CNC will not totally eliminate _____.

UNIT 5

Getting and Keeping the Job

KEY TERMS

sense of urgency   team player   cooperation
teamwork   appearance

PREPARING YOURSELF FOR A JOB

Companies only have jobs for one reason—to complete the tasks that help them make their products or provide their service in order to make money for the company. The job openings that companies have are therefore needs that must be filled to help the company get its tasks done and make money. In order to get a good job and fill any company's needs, you must do four things:

1. learn what those needs and tasks are,
2. learn what the skills are that will help meet those needs and tasks,
3. be genuinely interested in what you are trying to learn, and
4. set a goal to go about learning that particular set of skills, and focus on getting it done.

In Unit 4 we discussed what the jobs were that companies need to fill in the field of N/C and CNC and what basic skills were needed to fill those jobs. The remainder of this and other texts on N/C and CNC, in addition to shop "hands-on" training, will help you develop and learn the critical N/C and CNC skills. It's up to *you* to develop the interest, apply yourself, and get it done!

Getting the necessary training and skills for a job and convincing an employer that you have those skills is only half the battle. The other half is convincing the interviewer that you have the personal traits to make a good employee and that you will "fit in."

FIGURE 5–1
During the interview process, employers will try to learn many things about you. (Courtesy of Leo Burnett Inc., George Kerfren, photographer.)

In addition to determining whether you have all the skills required to perform the job, an employer will be trying to find out other things about you. Remember, you are competing against others who have similar backgrounds, experience, and skills. What do you offer an employer that sets you apart from the other candidates? Some other factors employers will be trying to find out about you during an interview (Figure 5–1) include: how dependable and reliable you are, your level of cooperation and *sense of urgency*, how eager, friendly, and enthusiastic you are, whether you are a *"team player,"* the impression you make on others based on your appearance and how you present yourself, and other factors. Let's talk about these so that you understand what an employer expects of you.

WHAT AN EMPLOYER IS LOOKING FOR

Dependability and reliability in people are just as important as in machines and cars. As we discussed in Unit 4, you depend on your car to get you to work and places you want to go. The manager who employs you depends on his equipment and *you* to make his products. If you or the equipment in his shop is unreliable or isn't dependable, then your employer doesn't make money and eventually you won't have a job and make money either. Your employer depends on you to be to work every day, to always be on

> **"Tech Talk"**
> In any job, punctuality (being on time) is critical to your success. Always leave enough time to get to work, school, or wherever you need to be, on time!

time, and to give him a "fair day's work for a fair day's pay." Dependability and reliability are very important to getting and keeping any job.

Your level of *cooperation* and *teamwork* is very important to your success in any job. Employers want people who work well with other people even though they may be of different ages, backgrounds, views, and cultures. Any employer will be trying to determine how well you will work with others, how cooperative you are, and whether you will be a "team player." Being able to cooperate and work successfully with other employees on a team (Figure 5–2) to achieve specific tasks and goals for the common good of the company is of major importance to both getting and keeping a job. In fact, teamwork is so important that many companies will first evaluate a person based solely on their ability to cooperate and work in small groups and teams with others. Only after it is determined that an individual is a "team player" will technical factors be looked at for employment consideration.

A "sense of urgency" is also important for you in getting and keeping a job. This refers to your ability to stay focused, react quickly, and to "move with urgency." Many people may be working but giving the appearance of not moving quickly and not caring whether the task they are working on gets done today or tomorrow. This

FIGURE 5–2
Being able to successfully work as part of a team may be very important to your success in any job or company. (Courtesy of Ford Motor Company.)

may leave management with the impression, right or wrong, that some workers are not doing their best and, therefore, are taking longer to accomplish things than others. Consequently, some people can become labeled as being "lazy" or "a loafer" whether it's true or not. On the other hand, workers can also be labeled "a good worker" or "a hustler." Make sure you develop that "sense of urgency" in all that you do and if any label is applied, it's that *you* are "a good worker" or "a hustler."

Friendliness and enthusiasm (Figure 5–3) also go a long way toward work relationships with your boss and your fellow workers. Always be friendly with everyone and work with enthusiasm on any task you are given. For sure, you will not like every job your

FIGURE 5–3
Friendliness and enthusiasm can go a long way toward getting and keeping a job.

boss gives you. But how you respond can go a long way to helping you keep your job or perhaps get a better one. When you are assigned something to do, make sure you do it with enthusiasm and to the best of your ability. And remember, do it with a smile.

Appearance is difficult to talk about and something that is generally a matter of opinion. Even though it is wrong, people form an opinion and impression about an individual based solely on how they are dressed. The only advice here is: be aware of how your co-workers dress and dress accordingly. Also, remember the rules of safety and make sure you dress appropriately for the work you are being asked to do.

We can talk about other traits and work habits such as honesty, personality, helpfulness and lots of other little things that count, but most of these are common sense. Just remember one simple rule: *you are the center of everything you do; be the best you can be, and do everything to the best of your ability.*

RECALLING THE FACTS

1. Dependability and reliability in _____ is just as important as in machines and cars.
2. Your level of _____ and _____ is very important to your success in any job.
3. Any employer will be trying to determine how well you work with others, how cooperative you are, and whether you will be a _____.
4. _____ refers to your ability to stay focused, react quickly, and to "move with urgency."
5. _____ and _____ goes a long way toward work relationships with your boss and your fellow workers.
6. Right or wrong, people will form an opinion and an impression of you based solely on your _____.

UNIT 6

Customer Focus and Importance

KEY TERMS

customer
customer-focused
external customer
customer-oriented
internal customer
value

DEFINING THE CUSTOMER

What or who is a customer? A *customer* is a person who buys anything (goods or services). Whenever you go to a store to spend your money to buy something (clothes, food, gasoline, go to a movie, get a haircut, etc.), you are a customer to that particular business, company, or store. What is it that pulls you or attracts you to that particular store or company as opposed to some other store or company? Why would you go to Wal-Mart for something instead of K-Mart, for example? Price? Location and convenience? Quality of the merchandise? The friendliness and helpfulness of the people working there? As a customer, you are free to spend your money wherever and however you choose. As a customer, you want *value* for your money. That is, you want the best and highest quality goods and services for the smallest amount of money. Businesses are constantly competing with one another for your dollars. To compete and be better than the competition, businesses are constantly striving to offer more and better quality products, improve their operations, reduce their costs, and increase their profits.

The company you work for or will work for has customers, too. If it didn't, you and your co-workers would not have jobs. You and everyone else in the company you work for has a responsibility to your customers. Customers expect value from your company for the product or service you and your co-workers provide, just as you expect value for the products you buy. If your customers can do better someplace else, they will switch and go to your competitor. Therefore, your goal and the goal of management and your co-workers is to continue to provide the best products and/or service at the lowest price and with the quickest delivery, and to continuously improve the company's products and processes. At the same time, the company must continually make a profit. After all, the goal of any business is to make money. Remember, however, just as you are a customer with things you, your friends, and family have bought, "A customer burned will never return."

The customers we have been talking about so far are *external customers*. External customers buy products or services from some other person or company, as seen in Figure 6–1. When you go

> **"Tech Talk"**
>
> A product's "life cycle" means the length of time a product stays on the market. Products today have increasingly shorter life cycles. Shorter life cycles mean that the first company that produces a product enjoys the highest profits associated with the first product introduction. As competition enters the market, the opportunity for all companies to make a large profit narrows.

FIGURE 6–1
As an external customer, you buy products or services from some other person, company, or store. (Courtesy of NCR Corporation.)

to Wal-Mart to buy a radio for example, you are an external customer to that store. But there are internal customers, too. *Internal customers* are people in your company, for example, who depend on you to do your job and complete your task so that they can do their job. Every activity and every job is a part of a process and everyone and every department in a company has a customer. For example, engineering's customer is manufacturing; manufacturing's customer is assembly, and so on. This is new thinking for some people because most have only thought of a "customer" in the external sense of the word.

KEEPING THE CUSTOMER SATISFIED

People working together must satisfy each other's needs as internal customers, as seen in Figure 6–2; otherwise, they won't be able to satisfy the needs of their external customers. This is necessary in order to bring order, harmony, and teamwork to the process, and why it's important to understand that everyone is a customer and has customers. Therefore, working in any job and for any company, you need to be customer-focused and customer-oriented in both the internal and external sense of the word.

What do *customer-focused* and *customer-oriented* really mean? Basically, they mean doing whatever it takes to fulfill the needs of your customers. And what are the needs of your customers? It is the things we have already talked about—cost, quality, reliability, service, support, on-time delivery, and, of course, value. For each person in an organization, it means doing your job, helping your co-workers (internal customers whose work efforts come before or after yours), and being a team player.

Some examples of being customer-focused and customer-oriented are:

- staying over a few minutes after quitting-time to help someone ship an item knowing that you're not getting paid and you would rather be on your way home
- asking someone with a critical task to be done and who needs help if you can help them when you already have too much to do

FIGURE 6–2
People who work together must satisfy the internal customers' needs in order to satisfy the needs of the external customer. (Courtesy Jervis B. Webb Company, Farmington Hills, Michigan.)

- making suggestions (verbal or written) to your supervisor to help improve product quality or process efficiencies in your department or in those of your co-workers
- participating on process improvement teams formed in your area or other areas
- going out of your way to track down a problem or identify its source instead of saying, "That's not my job"
- having a caring and conscientious attitude in all you do—it shows!

- taking the time to do it right instead of being in a hurry to "just get it done" figuring that someone else will fix anything that's wrong later
- working to identify a problem's cause in your own or a previous department instead of "just doing your job" and passing it on to the next person or department
- providing oral or written feedback to an internal or external customer about a problem's status (as opposed to telling the customer nothing, which is what he or she then assumes is happening with their problem—nothing!)

RECALLING THE FACTS

1. A _____ is a person who buys any goods or services.
2. As a _____, you want _____ for your money. That is, you want the highest quality goods and services for the smallest amount of money.
3. _____ _____ buy products or services from some other person or company.
4. _____ _____ are people in your company who depend on you to do your job and complete your task so that they can do their job.
5. Being customer-_____ and customer-_____ means doing whatever it takes to fulfill the needs of your customer.

UNIT 7

Continuous Improvement, Teams, and Your Involvement

KEY TERMS

continuous improvement	*teams*	*kaizen*
employee involvement	*value-added*	*waste*
non–value-added		

The areas of **continuous improvement, teams,** and **employee involvement** are part of almost any job today. Your participation in any of these types of programs or efforts is very important to any company you work for and for improving yourself. Let's spend some time discussing some important terms, concepts, and ideas.

THE PHILOSOPHY OF IMPROVEMENT

In order for businesses to remain competitive today, they must continuously improve. But what does this mean? Does it mean improve the quality of the products businesses produce? Does it mean improve the services businesses provide to customers? Does it mean improve the internal processes and procedures of a business and become more productive? The answer is: continuous improvement means all of these things—and more. It also means improvement of ourselves—to learn more, thereby increasing our value to ourselves, our companies, and the customers we serve. It means improvement in everything we do. Continuous improvement should know no boundaries.

Continuous improvement is rapidly gaining momentum in the United States, but our understanding of what it means is very narrow. When most people hear the words "continuous improvement" they immediately think of work: quality improvement programs (at work), employee involvement programs (at work), cost reduction programs (at work), and so on. The Japanese, however, have a much broader view of continuous improvement, and their own word for it: *kaizen*. Kaizen (pronounced "ki," like in kite, and "zan," like in van) takes on a much broader meaning than the narrow definition implied in the U.S. Kaizen means continuous improvement of one's personal life, home life, social life, and working life. In the workplace, kaizen means continuous improvement involving everyone (managers and workers) in everything.

The essence of kaizen is that it is deeply imbedded in the Japanese culture and mentality. Children are raised on it, adults live by it, and seniors teach and preach it. Kaizen is the basic philosophical foundation for the advancements in Japanese management and manufacturing techniques over the last 30 years. Kaizen is the key to Japanese competitive success; ongoing, continuous improvement involving everyone. It is everyone's job and everyone's business.

To the Japanese, kaizen is natural and obvious. Companies and plants do not remain the same for very long. To the Japanese, not a day should go by without some kind of improvement being made somewhere in the company. Ask any manager at a successful Japanese company what top management is pressing for and the answer will be kaizen.

In the workplace, kaizen starts with the recognition that any corporation has problems. Kaizen allows everyone to freely admit and flush all problems out into the open where they can be dealt with. Kaizen is customer-driven in cost, quality, and delivery; better products at lower prices.

The opposite of kaizen is keeping everything the same; status quo; ship tending; holding the course; no changes. Japanese visitors to some U.S. manufacturing plants who return five or more years later are amazed to find things exactly the same as they were during their last visit. It is inconceivable to Japanese management that a plant could stay the same for that period of time.

METHODS OF IMPROVEMENT

This difference in philosophy leads to some major differences in thinking about improvement in general between Japanese and U.S.

companies. Taking a look at Japanese versus U.S. methods of improvement you find some of the major differences listed below.

Japanese	*U.S.*
gradual change	abrupt change
long-term thinking	short-term thinking
incremental improvement	quarterly profitability
process improvement	innovation-results oriented

The Japanese are content to take a long-term view of change and improvement and make slow, gradual progress. In the U.S., short-term thinking and abrupt change are the norm, driven by the constant pressure of Wall Street and the stock market analysts for companies to show progressive increases in quarterly profits. This has a major impact on how companies view change, improvement, and the way companies measure and reward their chief executive officers.

Another difference is how Japanese versus U.S. companies view improvements and how improvements are made. Japanese companies, with their inbred culture of kaizen, tend to make a substantial, long-term commitment of time and effort to constantly improve internal processes through investing in people. U.S. companies, typically, don't want change to begin with. Change is usually forced on management by market conditions. The short-term pressures keep them heavily bottom-line and results-oriented, looking for new and fast innovations and quick-fix solutions through investment in new equipment and technology. Consequently, the simple, low-cost elements of improvement are largely ignored. U.S. companies also tend to favor equipment-based improvement versus the Japanese style of kaizen-based, operational improvement. Improvement with equipment:

> **"Tech Talk"**
> Quality is meeting the requirements and expectations of the customer, and is the least understood but most important factor in manufacturing and business survival.

- costs money
- takes time
- means that modifications are usually impossible
- is often not linked to cost reduction

Operational improvements, on the other hand:

- combine/consolidate work procedures
- change positions of objects, tools, benches, equipment, etc.
- install/modify simple tools, chutes, knock-out, and quick-change devices
- improve current equipment
- cost little or no money and are totally rooted in cost reduction

What this basically means is that companies need to complete the low-cost, things-you-can-easily-see-and-do, low-tech improvements before attempting to spend money on high-tech improvements like buying new equipment. Companies should improve through intelligence and ingenuity, finding simple ways to improve before spending money.

ELIMINATING WASTE

U.S. companies, however, are changing their attitudes toward their approach to continuous and operational improvement. As companies continue to streamline operations, downsize, cut costs, and improve profits, a way to provide low-cost, operational improvement is to eliminate non-value-added activities.

What are *non–value-added* activities and what are *value-added* activities? A value-added activity is one that shapes raw material or information to meet customer requirements. Examples would be the actual machining of a part (Figure 7–1), assembling finished parts, and answering or addressing customer questions and concerns. Non–value-added activities are those that take time, resources, or space but do not add to the value of the product itself. Examples would be parts sitting around waiting to be machined, excess inventory, a machine tool sitting idle while it is being set up, filling out unnecessary forms and paperwork (Figure 7–2), etc.

The essence of eliminating non–value-added activity is eliminating *waste* in manufacturing. But what is waste in manufacturing?

Waste in manufacturing consists of:

- defective products (scrapped parts and parts that need reworking)

FIGURE 7–1
The machining of a part is a value-added activity. (Courtesy—Jervis B. Webb Company, Farmington Hills, Michigan.)

- overproduction (producing too much)
- inventories (parts in process and finished sitting around)
- flow routes (distances parts have to travel to get worked on)
- processing (steps and paperwork required to get things done)
- transportation (actual movement of parts from place to place)
- waiting (machines waiting on parts, parts waiting on machines)

FIGURE 7–2
Filling out unnecessary forms and paperwork is an example of a non–value-added activity and is considered waste in manufacturing. (Courtesy of Levenger, *Tools for Serious Readers*.)

But what are the causes of waste in manufacturing? Although this list is by no means complete, a considerable amount of waste in manufacturing is caused by:

- layout distances (flow routes for part movement from machine to machine not laid out well, requiring parts to travel great distances)
- long setup time
- incapable manufacturing processes
- poor maintenance
- poor work procedures, plans, and methods
- lack of training or improper training
- lack of quality and maintaining quality standards

- poor supervision and management
- poor product design
- no performance standards or measurements
- ineffective production planning/scheduling
- poor equipment design, selection, or justification
- lack of workplace organization
- poor supplier quality and reliability

TEAMS

Another major focus and emphasis today is on teams. But what is a team? For starters, the company you work for or eventually will work for is a team united by the common bond of the company itself and dedicated to the common cause of making money for that company. Each individual in that company has a particular set of skills and a job to do in order for that team or company to be successful and make a profit. You have teams in sports. In baseball, football, basketball, or other team sports, each individual has his or her particular skill or talent and a job to do, just like in a company, and if all are successful, the team will be successful.

But teams in business today mean a lot more than just being part of the company team and using your particular education, training, and skills to do a job. It means working with other individuals across divisional, departmental, skill, education, and cultural boundaries on a much closer basis than ever before. It means recognizing that each person, because of his or her own special skills, talents, and experiences, has value and knows more about his or her own job than anyone else. And, he or she is recognized, consulted, and not criticized for that particular skill and knowledge.

In the past, management was content to make all the decisions and let the workers follow orders and do all the work. Now, management is beginning to recognize the value of each employee's knowledge about his or her particular job. Now the company wants your brain and your input on new products, new processes, new methods, and new procedures instead of just your muscle to work, to take directions, and to follow orders. As a result, you probably will be asked from time to time to participate on various teams because of some particular skill or knowledge that you have about the job you are or will be doing.

It is very important that you view these situations as opportunities and not just "something else to do." Someone thinks enough of your knowledge and ability about the job you do to seek your involvement and invite your participation, so make sure you see it as a positive opportunity for yourself and your company. Teams function to enhance a company, its organization, and its structure, not replace it.

Now let's talk a little about teams in general and your involvement. In general, teams can be formed by anyone for any purpose and for any period of time. Individuals are selected to participate on a team for two primary reasons:

1. they have a particular stake in the outcome of a problem or decision a team is trying to make and/or
2. they have a particular skill or knowledge that is required in order to help make a decision.

Typically, teams form, complete the project or task at hand that requires group participation, and disband.

Teams are formed in order to have a shared commitment to a purpose. They are always win-oriented and typically have or set specific goals and develop their own momentum as they progress. In some instances, completion deadlines may be imposed by management. Teams share group accountability (just like any sports team) and develop a common plan of action to accomplish their task. They are accountable to whomever commissioned them to form and assigned the task or goal.

There is much more to continuous improvement, teams, and employee involvement than the brief and simple overview discussed here. The intent here is to make you aware of some of the terms and programs used in industry today and to prepare you to recognize these in order for you to participate to the best of your ability. Teams and employee involvement are becoming increasingly important in business and industry in order for companies to gain competitive advantage. They will undoubtedly play a bigger and broader role in the future.

RECALLING THE FACTS

1. _____ improvement means improvement in everything. Continuous _____ knows no boundaries.
2. _____ is a Japanese word that means continuous improvement in one's personal life, home life, social life, and working life. In the workplace, _____ means continuous _____ involving everyone (managers and workers) in everything.
3. A _____ activity is one that shapes raw material or information to meet customer requirements.
4. _____ activities are those that take time, resources, or space but do not add to the value of the product itself.
5. The essence of eliminating any non–value-added activity is eliminating _____ in manufacturing.
6. _____ function to enhance a company, its organization, and its structure, not replace it. In addition, _____ are formed in order to have a shared commitment to a purpose and develop their own momentum as they progress.

UNIT 8

Math for CNC

KEY TERMS

seconds	line up decimal points	circle
pentagon	right angle	circumference
general/scalene	vertex	algebra
hypotenuse	minutes	trigonometry
diameter	quadrilateral	degrees
millimeter	decagon	obtuse
theorem	equilateral	triangle
angle	chord	octagon
acute	centimeter	radius
supplementary	Pythagorean	meter
polygon	geometry	ratio
hexagon	straight angle	right triangle
isosceles	complementary	

THE ROLE OF MATH IN CNC

Having a strong and solid math background is one of the most important requirements for being a good CNC programmer or operator. A programmer/operator will generally use a lot of basic math, such as addition, subtraction, multiplication, and division, in addition to some algebra, geometry, and a great deal of trigonometry. After all, numerical control is control of machine tools by *numbers*.

Modern CNCs today reduce the need to perform complex math calculations. Capabilities available on most controllers reduce the number of cutter location calculations required. And, many of the shop floor as well as off-line programming systems eliminate the need for manual calculations completely.

Although the advanced CNCs are capable of handling many of these calculations, the importance of math cannot be overstated. You not only use math to prepare new programs and modify existing ones, but also to check out completed programs before they are run on the machine. This is of critical importance because, ideally, you want to catch programming errors before they reach the machine. Some calculations, done before you run the program on the machine, can tell you whether your program looks like it will cut a good part and whether it will avoid wrecks and interferences with clamps, fixture components, or other parts of the machine.

For some, a discussion of the math topics in this unit will be just a review. For others, more time may be needed to learn some of the new material in this text or to obtain other sources for additional learning.

It is important to mention here that the math topics presented in this unit only cover certain scaled-down, CNC-related elements that should be of help to you as you progress with your CNC studies. It is not the intent to cover all aspects of each math topic at this time nor to suggest that this is all the math you'll need as you continue your study of CNC.

As you progress through this unit, reviewing and refreshing yourself with the CNC math concepts presented, think of math as another CNC "building block" of learning. There is hardly any career that does not involve some math. Being comfortable doing math is really no different than learning to speak well or write well. The more you know and the more you practice, the better and more comfortable you'll feel.

The list of CNC-related math topics to be discussed in this unit includes:

- addition, subtraction, multiplication, and division of decimals
- basic geometry
- metric conversion (inch to metric and metric to inch)
- introduction to algebra
- simple right triangle trigonometry

ADDITION, SUBTRACTION, MULTIPLICATION, AND DIVISION OF DECIMALS

In adding and subtracting decimals, the most important rule is to *line up decimal points*. The following examples show how the processes of addition and subtraction of decimals are performed and how to line up the decimal points.

Add: 7.837   Subtract: 456.218
        0.654   −129.126
      +43.528   327.092
       52.019

There are times when the decimals to be added or subtracted do not have the same number of decimal places. This condition does not present a problem because the "lining-up-the-decimal-point" rule still applies and missing digits can mentally be replaced with a zero.
Note the two following examples:

Add: 32.4   Subtract: 783.5
          3.152   −371.489
        250.   412.011
       + 43.07
        328.622

In multiplication, you multiply just as you do with whole numbers. The only difference is that you must be careful where you place the decimal point when you finish multiplying. For example, in multiplying 24.643 by 3.47 you get 85.51121. To place the decimal point, you count off as many decimal places to the right of the decimal point in the answer as there are decimal places to the right in both numbers.

If it should happen that a smaller number of digits appears in the answer than the number given by the rule above, enough zeros should be added to the right of the result to make up the proper number of digits. For example, in multiplying 0.25 by 0.38 you get an answer of 0.095, so another 0 must be added at the end of the number in order to provide four digits to the right of the decimal and arrive at .0950.

In dividing a decimal number by a whole number we proceed as in dividing with whole numbers and put the decimal point in the answer above its current location. For example $0.44/2 = .2$

In dividing a decimal by a decimal, as in the case of $20.64/0.8$, you move the decimal point one place to the right in both the dividend and divisor. This then makes the divisor a whole number similar to the case above ($20.64/0.8 = 25.8$).

BASIC GEOMETRY

The word *geometry* is made up of two words, "geo" meaning *earth* and "metron" meaning *measure*. The point, line, surface, and solid are the fundamental objects of geometry. Any one of the objects or any combination of them makes a geometric figure. The relative positions of the parts of a geometric figure determine its form or shape. Examples of geometric figures that you have probably already studied are the triangle, the rectangle, the cube, the circle, and others. Geometry is very important in your study of CNC because of the geometric figures (points, lines, circles, planes, etc.) and how they are used to describe and machine parts.

A very important concept in geometry is that of an *angle*. Angles are measured in *degrees*. Each hand of a clock, for example, passes through 360 degrees in one complete turn around the clock. At three o'clock, the hands of a clock form an angle of 90 degrees, or one-fourth of a complete turn. Such an angle is called a *right angle*. At 6 o'clock, the hands of a clock lie in a straight line and thus form an angle of 180 degrees, or one-half of a complete turn. Such an angle is called a *straight angle*.

In geometry, and in CNC, the positive rotation (turning) which forms an angle is thought of as opposite to that of the hands of a clock (counterclockwise). A negative rotation in geometry and in CNC, therefore, would be clockwise. When the rotation is less than 90 degrees, the angle is said to be *acute*; when it is more than 90 degrees, but less than 180 degrees, the angle is said to be *obtuse*. The point O (Figure 8–1) about which the rotation takes place and the

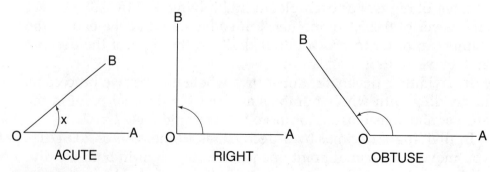

FIGURE 8–1
Some facts about angles.

point at which the sides of an angle come together is called the *vertex* of the angle. OA and OB are called the *sides*, or sometimes called the *arms* or *legs*.

Figure 8–2 illustrates some other important points about angles:

1. Two angles, like a and b in the first figure, which do not overlap and which have a common side and the same vertex, are called adjacent angles.
2. When two angles have the same vertex because two lines overlap, the angles are called *opposite angles* or *vertical angles*.
3. In geometry, angles can be added or subtracted. In the third figure, angle AOC is the sum of x and y.

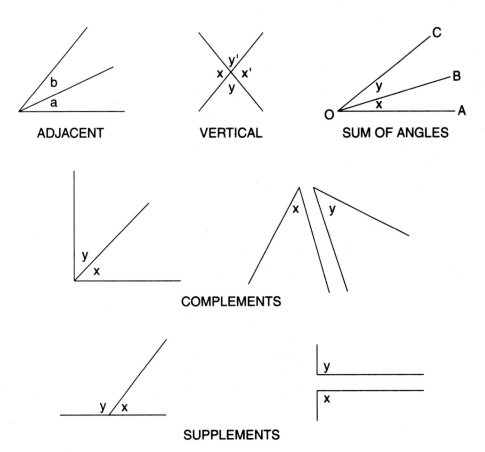

FIGURE 8–2
Some additional facts about angles.

4. If the sum of two angles is a right angle, as in the fourth group, each angle is considered the complement of the other, and the angles are said to be *complementary*.
5. If the sum of two angles is a straight angle, each is considered the supplement of the other, and the angles are said to be *supplementary*.

Since the degree of measurement does not give the accuracy required in all cases of angular measurement, subdivisions of degrees must be made. Each degree of measurement is divided into 60 equal parts called *minutes*, and each minute is divided into 60 equal parts called *seconds*. The symbol for the minute is ′ and the symbol for second is ″. Therefore, you have the following relationships:

$$60 \text{ seconds } (60″) = 1 \text{ minute } (1′)$$
$$60 \text{ minutes } (60′) = 1 \text{ degree}$$
$$90 \text{ degrees } = \text{right angle}$$
$$180 \text{ degrees } = 1 \text{ straight angle}$$
$$360 \text{ degrees } = 1 \text{ complete turn or revolution}$$

Sometimes the value of an angle must be changed from degrees, minutes, and seconds to decimal parts of a degree. To change 35 degrees, 19′ 12″ into decimal form, you first change 12″ to a decimal of minutes by dividing by 60. So, 12/60 = 0.2′ and, therefore, 35 degrees, 19′ 12″ = 35 degrees, 19.2′. Again, 19.2 can be changed to degrees by dividing 19.2′ by 60.

$$\frac{19.2}{60} = 0.32 \text{ degrees}$$

Therefore, 35 degrees, 19′ 12″ = 35.32 degrees.

It may also be necessary at times to convert the other way, from decimal parts of a degree to the degree-minute-second form. To do this it is only necessary to multiply the decimal degrees by 60 to convert to minutes and to multiply the decimal minutes by 60 to convert to seconds. For example, to convert 42.73 degrees to degrees, minutes, and seconds, first multiply .73 by 60 to get 43.8′ and then multiply .8′ by 60 to get 48″. Therefore, 42.73 degrees = 42 degrees 43′ 48″.

A *polygon* is a plane figure made up of three or more straight lines. The most common polygons are the *triangle* (three sides), the *quadrilateral* (four sides), the *pentagon* (five sides), the *hexagon* (six sides), the *octagon* (eight sides), and the *decagon* (ten sides), pictures of which are shown in Figure 8–3.

Because triangles are used so frequently in CNC, it is important to know their general classifications. The six figures shown in Figure 8–4 illustrate six different types of triangles classified on the basis of their sides and angles. The first three are classified with reference to their sides; the last three with reference to their angles. Important facts about triangles include:

- The name *general* (sometimes called *scalene*) is given to the triangle that has no two sides equal.
- The name *isosceles* (equal-legged) is given to the triangle that has two sides equal.

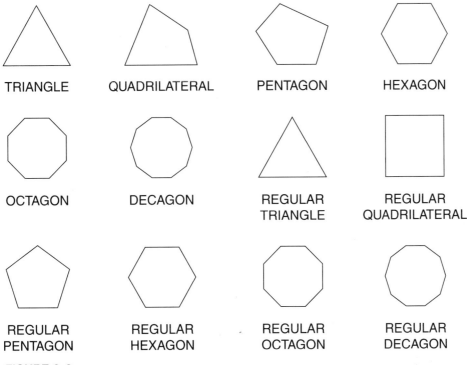

FIGURE 8–3
Some common polygons.

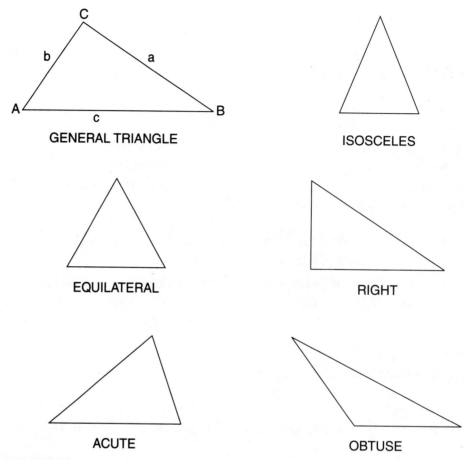

FIGURE 8-4
The general classifications of triangles.

- The name *equilateral* (equal-legged) is given to the triangle that has all three sides equal.
- A triangle in which all angles are acute is called an *acute* triangle.
- A triangle which contains one obtuse angle is called an *obtuse triangle*.
- A triangle that has one right angle (90 degrees) is called a *right triangle*. The *hypotenuse* is always the longest side of a right triangle and is always the side opposite the right angle. You

will learn more about right triangles in the discussion of right triangle trigonometry.

You are probably familiar with circles and have used and studied about them before. A definition of a *circle* is *a line drawn in a plane so that all points on the line are the same distance from a point in the middle called the center*. Some facts about a circle are:

- A straight line drawn from the center of a circle to the periphery of the circle is called the *radius*.
- A *chord* of a circle is a line connecting two points on the circle.
- The *diameter* of a circle is any straight line drawn through the center of the circle and ending at both ends of the circle. The diameter of a circle is the longest chord the circle can have. It is equal to twice the radius, or two radii.
- The distance around a circle is called the *circumference*.
- Part of a circle is called an *arc*.

METRIC CONVERSION

All new CNCs today are capable of working in the inch or metric system. CNCs are capable of sensing whether the input is inch or metric and adapting themselves accordingly to either inch or metric input.

Part drawings are either in inch or metric and some are dimensioned using both inch and metric so manufacturing people do not have to make metric-to-inch and inch-to-metric conversions. However, needs do arise for programmers and operators to make these conversions. Therefore, you need to know some basics of the metric system in order to make the conversions.

The basic unit of length in the metric system is the *meter*, which is equal to 39.37 inches. The meter is divided into 100 equal parts called *centimeters*, each of which is equal to .3937 inches. The meter can also be divided into 1,000 equal parts called *millimeters*, each of which is equal to .03937 inches. 10 millimeters = 1 centimeter. There are 25.4 millimeters in an inch.

The most common metric conversions required in shop applications are to change inches to millimeters and millimeters to inches. The conversions are:

$$\text{millimeters} = \text{inches} \times 25.4$$
$$\text{inches} = \text{millimeters} \times .03937$$

The conversion may also be expressed as:

$$\text{millimeters} = \text{inches}/.03937$$
$$\text{and}$$
$$\text{inches} = \text{mm}/25.4$$

It is important to have an approximate idea of what the answer should be before the conversion calculation is made. For example, the millimeter equivalent is larger (by 25.4 times) than the inch equivalent.

INTRODUCTION TO ALGEBRA

Occasionally, CNC programmers/operators must use simple algebra and formulas to solve a problem. *Algebra* is an extension of arithmetic that makes use of letters and symbols as well as numbers. In order to see the advantage of algebra over arithmetic, you should first compare a common arithmetic rule with its algebraic formula.

Arithmetic	*Algebra*
1. If oranges cost 7 cents each, 4 oranges cost 4 × 7 cents.	1. If oranges cost P cents each, N oranges cost $N \times P$ cents, or NP cents.
2. The cost of any number of oranges is the cost of one orange multiplied by the number of oranges.	2. If C represents the total cost in cents, then $C = NP$.

The equation $C = NP$, which expresses the arithmetic rule for finding the cost of oranges, is called a *formula*. Such a formula is often referred to as a *cost formula*.

You can see that an algebraic formula is much shorter than its corresponding arithmetic rule. Formulas are widely used in arithmetic, algebra, geometry, science, and the shop. In writing formulas you use letters.

Solving a simple formula or equation means determining the value of the unknown or unknowns that will make the equation true. This solution may be accomplished in one of several ways. We may divide both sides of the equation by the same number, multiply both sides by the same number, subtract the same number from both sides, or add the same number to both sides. In any one of these four cases, the "balance" of the equation is not changed, but the process is intended to leave the unknown value alone on the left side of the equal sign.

Some examples include:

a. Dividing both sides of the equation $4n = 20$ by 4 gives the new equation

$$\frac{4n}{4} = \frac{20}{4} \text{ or } n = 5$$

b. Multiplying both sides of the equation $1/2n = 6$ by 2 results in the new equation $(1/2n)(2) = (6)(2)$ or $n = 12$.

c. Subtracting 3 from both sides of the equation $n + 3 = 9$ results in the new equation $n + 3 - 3 = 9 - 3$ or $n = 6$.

d. Adding 2 to both sides of the equation $n - 2 = 8$ results in the equation $n - 2 + 2 = 8 + 2$ or $n = 10$.

   Sometimes it is necessary to combine two or more of the above operations to solve the equation.

e. In the equation $2/3n - 2 = 6$, you multiply both sides by 3 giving $2n - 6 = 18$, then divide by 2 giving $n - 3 = 9$, and finally add 3 to both sides giving $n - 3 + 3 = 9 + 3$ or $n = 12$.

SIMPLE RIGHT TRIANGLE TRIGONOMETRY

To begin your study of trigonometry, you first need to study the basic right triangle. This involves learning the Pythagorean theorem. The *Pythagorean theorem* is used to calculate the length of an un-

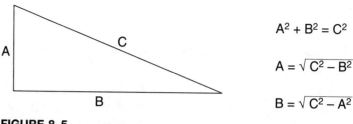

FIGURE 8–5
The Pythagorean theorem.

known side of a right triangle if the lengths of the other two sides are known. This theorem or formula states that if you square each of the legs of a right triangle and add the results together, it will equal the square of the longest side, called the *hypotenuse*. The formula is shown in several different forms in Figure 8–5.

Right triangle *trigonometry* is used to find unknown sides and angles in cases where at least one side and one angle are known. Trigonometry is one of the most useful forms of mathematics and is used extensively in shop applications. Trigonometry, which comes from Greek words meaning "triangle measure," is basically the study of angles.

It is customary in trigonometry to use a standard right triangle, lettered as those shown in Figure 8–6. Study the examples and formulas in Figure 8–6.

Values are assigned to these formula *ratios* and stored in tables or the values can be determined through the sine, cosine, or tangent functions on a technical calculator. You must understand that the trigonometric ratios do not indicate inches, feet, degrees, or measurements of any kind. They are pure ratios, and are, therefore, pure numbers without any units assigned to them.

The numerical values of the trigonometric ratios are for any angle and do not depend on the lengths of the sides of the angle but only on the size of the angle. For this reason, you can say that trigonometric ratios are functions of the angle. The numerical values of all trigonometric ratios have been computed for angles for zero to 90 degrees. These values are usually in tables and the values are given for tenths of degrees or for minutes.

Some sample problems with solutions can be seen in Figure 8–7.

The study of trigonometry involves ratios (a ratio is a relationship between two similar things) of angles and sides of right triangles. Virtually all right triangle trigonometry needed for CNC programming is expressed in the following formulas:

$$\text{sine} = \frac{\text{side opposite}}{\text{hypotenuse}} \quad \text{cosine} = \frac{\text{side adjacent}}{\text{hypotenuse}} \quad \text{tangent} = \frac{\text{side opposite}}{\text{side adjacent}}$$

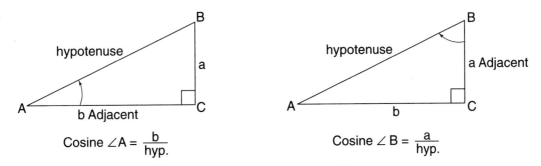

The sine function (pronounced "sign" like in "stop sign") is the ratio of the opposite side of a right triangle to its hypotenuse for a given angle. Determine the side opposite based on the angle referenced as in the example above.

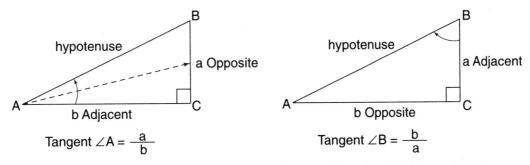

The cosine function is the ratio of the adjacent side of a right triangle to its hypotenuse for a given angle. Determine the side adjacent based on the angle referenced as in the example above.

The tangent function is the ratio of the opposite and adjacent sides of a right triangle for a given angle. Determine the side opposite and side adjacent based on the angle referenced as the example above.

FIGURE 8–6
Standard trigonometry formulas and triangles.

76 • UNIT 8

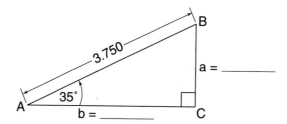

In this example, one angle and the hypotenuse are given. Find side a + b.

Step one: Find the size of the third angle, angle B. Remember, the sum of the angles of a triangle = 180°. ∠C = 90°.
∠A + ∠B = 90
∠B = 90° − ∠A
∠B = 90° − 35°
∠B = 55°

Step two: Find the value of side a by using the sine function. This function uses two known values (an angle and the hypotenuse) to find the third unknown (side opposite).

$$\text{Sine} = \frac{\text{side oposite}}{\text{hypotenuse}} \quad \text{(write the formula)}$$

$$\text{Sine } \angle A = \frac{a}{3.750} \quad \text{(set up the problem)}$$

Sine ∠A = 35° = 0.57358 (find in trig table or use calculator)

$$0.57358 = \frac{a}{3.750}$$

$$3.750 \times 0.57358 = \frac{a}{\cancel{3.750}} \times \frac{\cancel{3.750}}{1} \quad \text{(cross multiply)}$$

2.15 = a   or   side a = 2.15

Step three: Find the value of side b by using the cosine function. This function uses two known values (an angle and the hypotenuse) to find the third unknown (side adjacent).

$$\text{Cosine} = \frac{\text{side adjacent}}{\text{hypotenuse}} \quad \text{(write the formula)}$$

$$\text{Cosine } \angle A = \frac{b}{3.750} \quad \text{(set up the problem)}$$

Cosine ∠A = 35° = .81915 (find in trig table or use calculator)

$$.81915 = \frac{b}{3.750}$$

$$3.750 \times .81915 = \frac{b}{\cancel{3.750}} \times \frac{\cancel{3.750}}{1} \quad \text{(cross multiply)}$$

3.07 = b   or   side b = 3.07

FIGURE 8–7
Sample problems and solutions.

Example 2

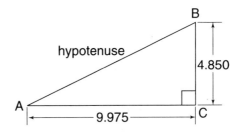

In this example, 2 sides of the triangle are given. Find ∠A, ∠B, and side AB (hypotenuse).

Step one: Find ∠A. Use the tangent function because it uses two known values (two sides) to find the third unknown value (angle A).

$$\text{Tangent} = \frac{\text{side opposite}}{\text{side adjacent}} \quad \text{(write the formula)}$$

$$\text{Tangent } \angle A = \frac{4.850}{9.975} \quad \text{(set up the problem)}$$

Tangent ∠B = .48621   (find this value in trig table or use a technical calculator)

∠A = 25° 56′

If ∠A = 25° 56′, to find ∠B you subtract 25° 56′ from 89° 60′ (which is the same as 90°) to find the complement (89° 60′ − 25° 56′ = 64° 04′).

Step two: Find side AB or the hypotenuse. Use the sine function; angle A (and angle B) and the side opposite are known.

$$\text{Sine} = \frac{\text{side opposite}}{\text{hypotenuse}} \quad \text{(write the formula)}$$

$$\text{Sine } 25° 56′ = \frac{4.850}{AB} \quad \text{(set up the problem)}$$

$$\text{Sine } 25° 56′ = .4373 = \frac{4.850}{AB} \times AB \quad \text{(cross multiply)}$$

$$= \frac{4.850}{.4373}$$

$$AB = \frac{4.850}{.4373}$$

AB (hypotenuse) = 11.0907

FIGURE 8–7
Continued

RECALLING THE FACTS

1. The most important rule to remember in adding and subtracting decimal numbers is to _____.
2. The point, line, surface, and solid are the fundamental objects of _____.
3. Angles are measured in _____, _____, and _____.
4. An angle of 90 degrees is called a _____ *angle* while an angle of _____ degrees is called a *straight angle*.
5. An angle of less than 90 degrees is called an _____ *angle*; an angle of greater than 90 degrees is called an _____ *angle*; and where the two sides of an angle come together is called the _____.
6. If the sum of two angles is 90 degrees, a _____ angle, the angles are said to be _____. If the sum of two angles is a straight angle, _____ degrees, the two angles are said to be _____.
7. A _____ is a plane figure made up of three or more sides. Name the _____ below by the number of sides the figure has.

    three sides _____
    four sides _____
    five sides _____
    six sides _____
    eight sides _____
    ten sides _____

8. The two names given to the triangle that has no two sides of equal length are _____ and _____.
9. A triangle that has two sides equal is called an _____ triangle and one that has three sides equal is called an _____ triangle.
10. A triangle that has one right angle (_____ degrees) is called a _____ _____. The longest side of that triangle and the side that is always opposite the right angle is called the _____.

11. The definition of a _____ is a line drawn in a plane so that all points on the line are equidistant from a point in the middle called the center.
12. A straight line drawn from the center of the _____ to the _____ is called the _____; the _____ of a _____ is a line drawn through the center of the _____ and ending at both ends of the _____; a _____ is a line connecting any two points on the _____.
13. The basic unit of length in the metric system is called the _____, which is _____ inches.
14. The _____ is divided into 100 equal parts called _____; each _____ is equal to _____ inches.
15. The _____ is divided into 1,000 equal parts called _____; each _____ is equal to _____ inches.
16. _____ is an extension of arithmetic that makes use of letters and symbols as well as numbers.
17. The _____ is used to calculate the length of an unknown side of a right triangle if the lengths of the two other sides are known.
18. A _____ is the relationship of one item to another.
19. _____ is basically the study of angles.

UNIT 9

Practicing Your Math

KEY TERMS

perpendicular   *parallel*   *intersect*
bisector

In this unit you will practice the math you reviewed and learned in Unit 8. This entire unit is designed to give you practice and build your confidence with doing some math problems similar to those that you would do in the shop.

Some of the math problems in this unit are just simple practice problems, so you won't even need your calculator for these. Others are designed to make you think. Check all your work because math mistakes with CNC can be very damaging and costly, as you already know. Hopefully, you will enjoy doing all of these and become more comfortable with solving these types of problems as you learn more about CNC programming and operation.

Becoming capable, confident, and comfortable doing math is very important to CNC programmers and operators because without basic math skills, programmers and operators cannot prepare their work, check their work, or avoid serious accidents or injury.

First, let's start with some simple exercises followed by some addition, subtraction, multiplication, and division problems.

Write in the figures for each of the following numbers:

1. Eight hundred twenty-seven thousandths _____
2. Seven ten-thousandths _____
3. One hundred five thousandths _____
4. Twenty-five and thirty-eight thousandths _____
5. Thirty-five thousandths + eighteen thousandths _____

Add:

6.	7.6 12.2 16.4	7.	17.8 8.2 36.	8.	19.5 0.9 74.1	9.	16.22 .09 56.73	10.	28.77 57.80 3.48
11.	7.034 5.006 9.765	12.	156 87.29 385.42	13.	67.128 43.095 6.867	14.	375.6 742.3 487.4	15.	0.769 4.087 3.605

16. 28.7, 128.6, 0.7, 230 = _____
17. 6.54, .008, 12.1, 107.3 = _____
18. 12.57, 6.2, .054, 8, 123.9 = _____

Subtract:

19.	57.6 23.5	20.	59.46 41.03	21.	7.88 5.60	22.	1.475 0.43	23.	26.87 3.42
24.	8.5 4.73	25.	36.2 12.637	26.	87.55 24.48	27.	94.57 67.8	28.	6. 2.5

29. 28.4 − 18.9 = _____
30. 37.243 − 1.927 = _____

Multiply:

31.	6.8 5.6	32.	58 4.5	37.	12.9 6.3	34.	5.28 2.07	35.	0.33 2.70
36.	12.62 10	37.	4.2 .7	38.	13.6 .007	39.	.043 .13	40.	18 2.02

41. 18 × .02 × .007 = _____
42. .008 × 2.4 × .6 = _____

Divide:

43. $0.05\overline{)17.65}$
44. $0.80\overline{)287.28}$
45. $9.0\overline{)4.3758}$
46. $40\overline{)33.408}$
47. $300\overline{)2769.9}$
48. $0.70\overline{)6.671}$
49. $0.6\overline{)0.43812}$
50. $2.0\overline{)1.010}$
51. $8.6\overline{)3749.6}$
52. $10.6248 \div 93.2 = $ _____
53. $86 \div 0.4 = $ _____
54. $375 \div 0.25 = $ _____
55. A 90° angle is called a _____ angle.
56. The point at which angular rotation takes place is called the _____ .
57. When an angle is less than 90°, it is called an _____ angle. If the angle is greater than 90°, it is called an _____ angle.
58. All angles of a triangle when added together must equal _____ degrees.
59. If one angle of right triangle is 58° 32′, the remaining angle is _____ .
60. The _____ of a circle is equal to twice the radius; the distance around the circle is called the _____ .
61. Change 20° 24′ 36″ to decimal form.

    _____

62. Change 31° 45′ 45″ to decimal form.

    _____

63. Change 12.71° to degrees, minutes, seconds.

    _____

64. Change 53.23° to degrees, minutes, seconds.

    _____

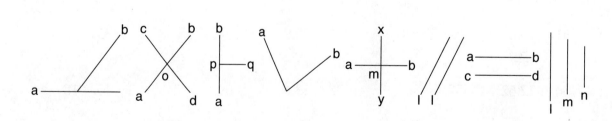

65. When two lines meet or intersect each other as shown in the first two figures in Problem 64, they form _____ with each other.

66. If two lines meet or *intersect* each other (come together) at right angles, as shown in the middle group of three figures in Figure 9–1, the lines are said to be *perpendicular*. The symbol for "perpendicular" or "is perpendicular to" is ⊥. Therefore PQ ⊥ AB is read PQ is _____ to AB; and a ⊥ b is read a is _____ to b.

67. In the middle group of figures in Figure 9–1, the line XY divides AB at M into two _____ parts; that is _____ = _____. The line XY is the *bisector* (the line that splits another line into two equal parts) of _____, and since XY ⊥ AB, it is called the *perpendicular bisector* of AB.

68. In geometry, lines in the same plane which do not _____, however far they are extended, as seen in the last group of three figures in Figure 9–1, are *parallel*. The symbol for "_____" or "is parallel to" is ∥.

69. The statement AB∥CD is read, AB is _____ to CD.

70. Lines L, M, and N are _____ to each other.

71. Any line which divides a right angle forms two angles which are adjacent _____ angles.

72. The supplement of 56° 25′ is an angle of _____° _____′ and the complement of the same angle is _____° _____′.

Match the angles and figures for questions 73–82.

73. Pentagon _____

(a)

74. Obtuse _____

(b)

75. Parallel lines _____

(c)

76. Perpendicular _____

(d)

77. Adjacent angles _____

(e)

78. Isosceles _____

79. Hexagon _____

80. Complementary angles _____

81. Vertical angles _____

82. Supplementary angles _____

(f)
(g) 59°12' 120°48'
(h)
(i)
(j) A C B D

83. Convert 26 inches to mm _____
84. Convert 78.5 inches to mm _____
85. Convert 1.001 inches to mm _____
86. Convert 56 mm to inches _____
87. Convert 35 mm to inches _____
88. Convert 9.525 mm to inches _____

Solve each of the following equations.

89. $7N = 28$ N = _____
90. $48 = 16T$ T = _____
91. $8A = 40$ A = _____
92. $27 = 9K$ K = _____
93. $30A = 20$ A = _____
94. If you divide both sides of the equation $1/2\ H = 3$ by $1/2$ you obtain the equation _____ = _____. Since dividing by $1/2$ is the same as multiplying by _____ you can arrive at the same result by _____ both sides of the equation by _____.
95. In order to solve the equation $1/4\ C = 8$, you must _____ both sides of the equation by _____. You arrive at the conclusion that _____ = _____ and this is the correct result because $1/4 \times$ _____ = _____.

Match the algebraic formulas on the left with the corresponding statements on the right by writing the statement letter on the line provided that matches the algebraic translation of the statement.

96. _____ $3N - 4 = 14$ (a) One-half a certain number is equal to the number itself decreased by 14.

97. _____ $N - 14 = 1/3\,N$ (b) When 24 is decreased by twice a certain number, the result is 14.

98. _____ $N + 9 = 14$ (c) A certain number decreased by 5 is 14.

99. _____ $N + 6 = 14$ (d) When 9 is added to a certain number, the result is 14.

100. _____ $1/2\,N = N - 14$ (e) When a certain number is increased by 6, the result is 14.

101. _____ $N - 5 = 14$ (f) When 14 is subtracted from a certain number, one-third of the number remains.

102. _____ $24 - 2N = 14$ (g) Three times a certain number decreased by 4 is 14.

Find Side AB and BC.

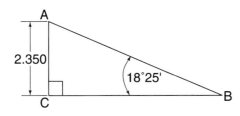

Step 1: Write the formula for finding AB.

Step 2: Set up the problem.

Step 3: Solve.

103. Side AB = _____

　　Step 1: Write the formula for finding BC.

　　Step 2: Set up the problem.

　　Step 3: Solve.

104. Side BC = _____
　　Find the hypotenuse.

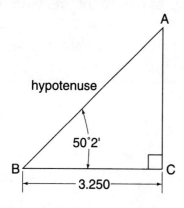

　　Step 1: Write the formula for finding the hypotenuse.

　　Step 2: Set up the problem.

　　Step 3: Solve.

105. Hypotenuse = _____
 Find side X.

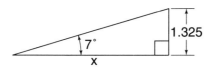

Step 1: Write the formula for finding side X.

Step 2: Set up the problem.

Step 3: Solve.

106. Side X = _____
 Find ∠A, ∠B, and side Y.

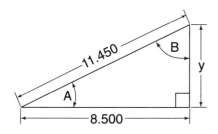

Step 1: Write the formula for finding ∠A.

Step 2: Set up the problem.

Step 3: Solve.

107. ∠A = _____
108. ∠B = _____

Step 1: Write the formula for finding side Y.

Step 2: Set up the problem.

Step 3: Solve.

109. Side Y = _____
Find angle X.

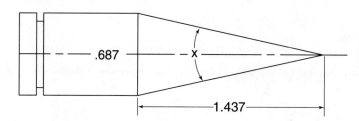

Hints:
- Find the right angle.
- Take $1/2$ the diameter.
- Don't forget to double the angle; you're only finding one-half of it as part of the triangle.

Step 1: Write the formula.

Step 2: Set up the problem.

Step 3: Solve.

110. ∠X =
    Find ∠A.

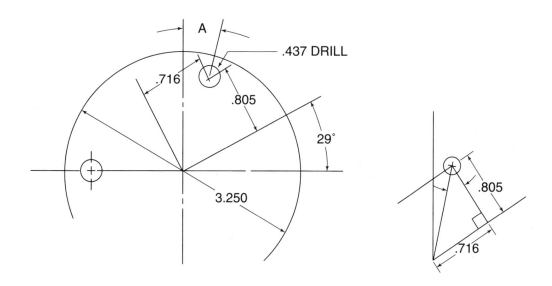

Hints:
- Sketch lines to find triangle and right angle.
- Find ∠ opposite .716 dimension in sketch.
- Subtract from 90° to find complementary angle.
    Step 1:   Write the formula.

    Step 2:   Set up the problem.

    Step 3:   Solve.

111. ∠A = _____

RECALLING THE FACTS

1. If two lines meet or come together, they are said to _____.
2. When two lines _____ at right angles to each other, they are said to be _____.
3. A line that splits another line into two equal parts is called a _____.
4. In geometry, lines in the same plane which do not _____, however far they are extended, are said to be _____.
5. ⊥ is the symbol for _____ and ∥ is the symbol for _____.
6. The longest side of a right triangle and the side that is always opposite the right angle is called the _____.
7. Fill in the following trig functions:

    $$\underline{} = \frac{\text{side opposite}}{\text{hypotenuse}}$$

    $$\underline{} = \frac{\text{side adjacent}}{\text{hypotenuse}}$$

    $$\underline{} = \frac{\text{side opposite}}{\text{side adjacent}}$$

8. An angle is 35 degrees; its complement is _____; and its supplement is _____.

SECTION 1
REVIEW

KEY POINTS

- N/C and CNC machines have no hand wheels or levers; they have electronic controls attached to them which guide and direct the machine tools.
- N/C and CNC machine tools are a lot like people; alike in some ways and different in others.
- The first and most important thing to learn about N/C and CNC machine tools is *safety*.
- With N/C and CNC, machine moves are typically thought out and assembled into a "package" called a *program*. *You* are still responsible for safety whether you wrote the program or operate the machine.
- Pay attention to all safety rules and always know where the "master stop" button is located on all machinery before you operate it.
- Although John Parsons came up with the idea of numerical control, MIT came up with the name.

I learned a lot in this section!
Let's review!

- There are many different types, styles, and ages of N/C and CNC machines producing parts in manufacturing plants.
- No greater change has occurred in the field of N/C than with the machine control unit (MCU); MCUs have progressed from bulky tube types to the microprocessor-based units of today.
- MCU capability and reliability have increased as size and cost have decreased.
- Hard-wired controls (N/Cs—built up to the early 1970s) have all the logic for the codes, commands, and functions "wired-in" and fixed within the control unit. Software-based controls (CNCs—built after the early 1970s) contain microprocessors, have all the logic for the codes, commands, and functions loaded in the computer logic of the CNC, and the logic can be easily changed.
- The executive program or load tape gives the control unit its "brains" and makes it "think" like a machining center or lathe; it resides within the CNC and can be altered or changed to add new functions.
- CNC machines have become easier to use because of advanced electronics, menu-selectable displays, advanced graphics, and conversational English programming.
- People with N/C and CNC experience have been and will continue to be in demand; this is because CNC machines are the relatively new "tools" of manufacturing.
- The typical N/C and CNC jobs associated with most manufacturing companies are: operator, operator/programmer, setup person, programmer, tooling engineer, and methods engineer.
- Three things that N/C and CNC machines can do better than manual machines are: increase accuracy, improve repeatability, and provide higher reliability.
- Getting the necessary training and skills for a job and convincing an employer that you have those skills is only half the battle; the other half is convincing the interviewer that you have the personal traits to make a good employee and that you will "fit in."
- CNC will not totally eliminate errors.

- Customers are very important to the success of any business. Customers are both internal and external.
- Kaizen is continuous improvement of one's personal life, home life, social life and working life. In the workplace, kaizen means continuous improvement involving everyone, in everything.
- The opposite of kaizen is complacency; status quo; keeping everything the same; holding the course; no changes.
- A value-added activity is one that changes raw material or information to meet customer requirements. Non–value-added activities are those that take time, resources, or space but do not add to the value of the product itself.
- Teamwork in business today means working with other individuals across divisional, departmental, skill, education, and cultural boundaries to achieve a common purpose.

REVIEW QUESTIONS AND EXERCISES

1. The first and most important thing to learn about N/C and CNC machines is _____.
2. With N/C and CNC machines, cutting action and machine moves are thought out and assembled into a "package" called a _____.
3. If you are operating an N/C or CNC machine tool, you are responsible for *all* of its actions whether you programmed the machine or not. (True/False)
4. Circle safety-related problems in the following list:
    a. visitors without safety glasses
    b. operator wearing wristwatch
    c. cutting tools lying on shop rag on top of work bench
    d. machine waiting to be unloaded
    e. dial calipers lying on machine table
5. Compressed air can be used to clean machine surfaces, cabinets, controls, and the general work area. (True/False)
6. Gloves or a shop rag should only be used when changing tools by hand. (True/False)

7. Machine safety guards can only be removed if the cutting action is difficult to see. (True/False)
8. In terms of safety around N/C and CNC machines, the "master stop" button is the most important button on the control.
9. The driving force behind the introduction of N/C was:
   a. The parts for U.S. planes and missiles were becoming too complex to be made on manual machines.
   b. MIT and Parsons needed a research project.
   c. Manual machines weren't reliable anymore.
   d. Computers were a good thing to start using.
10. The greatest changes in the field of N/C has occurred with the _____.
11. As advanced electronics began to be used in CNCs, control size decreased but cost of the control units increased. (True/False)
12. _____ controls contain all fixed logic; _____ controls contain changeable logic.
13. New codes, commands, or functions cannot be added to a CNC's executive program. (True/False).
14. The CNC manufacturer is responsible for preparing and loading the executive program into the CNC. (True/False)
15. Explain why people with N/C and CNC experience have been, and will continue to be, in demand.
16. The job of an _____ is an entry-level position in the field of N/C machining that generally involves loading and unloading parts but still allows the opportunity to learn a great deal about N/C machining.
17. List five reasons why N/C and CNC machining are better than manual machining.
18. The four things you must do to get a good job and fill a company's needs are: (1) learn what those needs and tasks are, (2) learn what the skills are that will help meet those needs and tasks, (3) be genuinely interested in what you are trying to learn and, (4) _____.
19. Dependability and reliability in _____ is just as important as in machines and cars.

20. List and explain the meaning of five new words in this section.
21. Explain what kinds of changes can be made in a shop to provide operational improvements.
22. Explain the difference between a value-added and a non–value-added activity. Give examples of each.
23. List five examples of waste and five causes of waste in manufacturing.
24. Using full sentences, write a paragraph discussing the importance and value of teams in business today.

SHOP ACTIVITIES

1. Go out into the shop and identify N/C or CNC machine tools. Walk around and look closely at them. Are they all the same? What are the visual differences? Can you identify the readout screen, keyboard, function keys, etc. on the control?
2. Identify "master stop" buttons on the controls. Would they be easy to see and get to if you were operating the control?
3. Walk around the shop. Try to identify as many safety problems as you can find. Make a list. What would you do to correct them? Give the list to your instructor or supervisor.
4. Pretend for a moment that you owned a shop and were going to hire someone for an N/C or CNC job. Make a list of five questions that you would like to ask that person when you interviewed him or her. Other than what was discussed in the text, what other personal traits would you be looking for in a person?
5. Form a team of a few people to identify some non–value-added activities and areas of waste in your shop. Make a list of these and, with your team, meet with your supervisor/instructor to discuss these, their causes, and what, if anything, can be done to eliminate or reduce them.

SECTION 2

CNC OPERATION AND COMMUNICATION

OBJECTIVES

After studying this section, you will be able to:

- Explain how N/C data is stored and used.
- Describe axis relationships, motion, and feedback control.
- Explain the types of N/C and CNC systems.
- Name the types of N/C data input and coding formats.
- Identify the causes and prevention of N/C and CNC downtime.

How does CNC operate?

INTRODUCTION

This is a very important section because it discusses how CNC machines operate and how information is communicated between the machine and the MCU.

Specifically, you will learn how N/C data is stored and used. As this data is read, stored, and acted upon, it causes machine action and motion through the machine's drive motors. You will learn about the different types of N/C and CNC machines along with the axis relationships that form the basis for all programming and machine axis movement. The different types of N/C data input and storage will also be discussed, along with what causes MCU downtime.

It is important to know how CNC machines operate for two reasons. First, if you don't know what the machine does, how it operates, and the principles upon which programming is based, you can't visualize the cutting action, which is so important to successful programming. (We discussed the importance of "visualizing" the cutting action when programming was discussed in Section 1.) Second, knowing how CNC machines operate and the means to communicate with them is important for safety reasons. Trying to program or operate a CNC machine without knowing what will happen and what the machine will do can cause serious injury to yourself or your co-workers and cause damage to the machine, the part, and the cutting tool. Knowing and understanding CNC operation and communication is a very important step in building your knowledge base in this field.

UNIT 10

How N/C Data is Stored and Used

KEY TERMS

machine actuation registers
buffer storage

PROGRAMMING THE MACHINE

With an N/C or CNC machine, the sequence of events starts with how a part will be manufactured, as seen in Figure 10–1. Manufacturing and manfacturing engineering personnel, including the part

FIGURE 10–1
N/C and CNC machines must be programmed before the machine can be run. The process begins by studying the part print and sometimes consulting with peers. (Courtesy of Cincinnati Milacron, Inc.)

programmer, sometimes discuss a workpiece part drawing to determine how to make the part in the quickest, safest, and most economical way. Some of the questions a part programmer asks before beginning to program are:

- What operations are required?
- What machines will be used?
- How will the part be held? Will it be held in a vise or in a fixture?
- What operation needs to be done first, second, third, etc.?
- What tools are needed and in what order should they be used?

The part programmer must answer these questions whether the CNC machine is programmed off-line or at the machine. Instead of turning hand wheels and proceeding by touch, experience, and know-how, as a manual machine operator would, the part programmer "visualizes" the steps and processes and the machine motions necessary to machine the part.

The programmer then proceeds to document the logical steps necessary in order to machine the part. This is done whether the program will be prepared manually at the machine or with the assistance of a computer. The manuscript data is then converted to the language of the MCU for the machine tool on which the part will be machined.

N/C data can enter the machine control unit through one or more of five means:

1. tape readers
2. cassette tapes
3. floppy disks
4. direct line to the CNC
5. manual data input (MDI)

Each of these will be discussed in more detail later in this section.

USING THE DATA

Once inside the MCU, N/C data is then passed to *machine actuation registers* inside the control. These registers are like storage mailboxes within the control that accept the N/C data consisting of machine coordinates and other codes, commands, and functions. From the machine actuation registers, the N/C data is then decoded and passed to the machine tool drive motors to move and position the machine slides and table, start and stop the spindle, and other machine commands. A drawing of this process is shown in Figure 10–2.

> **"Tech Talk"**
>
> Computers are simply tools that extend the effectiveness of man's brain the way the machine tool extends the effectiveness of man's muscle.

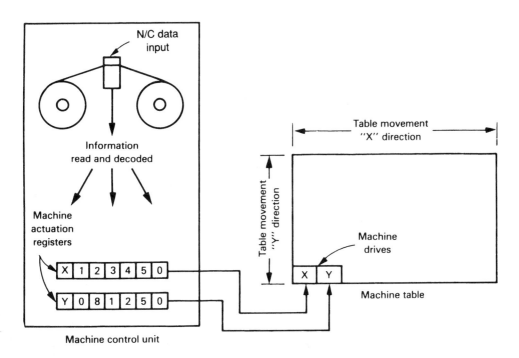

FIGURE 10–2
Basic sketch of N/C data input being read, decoded, and passed to machine actuation registers inside the control causing movement in the motors, thereby fulfilling machine commands.

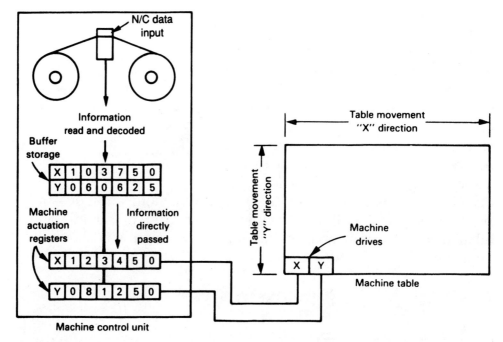

FIGURE 10–3
Basic sketch of N/C data input information being read, decoded, and held in buffer storage until machine actuation registers have completed previous operations and commands. Information is then passed from buffer storage to actuation registers.

Today, CNC controls are equipped with *buffer storage.* As shown in Figure 10–3, this feature allows information to be stored in temporary "holding" registers while an operation is being performed from the active machine registers. When the machine is finished with the N/C data in the active machine registers, the next set of data stored in the buffer registers is transferred instantaneously to the active machine registers. The advantage of having buffer storage in a CNC control is that it shortens the time between reading the N/C data and the machine acting on the data. Basically, this means that the transfer of N/C data from the buffer storage to the active machine registers is faster than reading the N/C data directly into the control unit, decoding it, and sending it directly to the machine actuation registers.

Buffer storage reduces the dwell time between machine moves. This is because the next N/C instruction is read and stored while the machine is acting on the previous instruction. Part finish is improved because the machine does not have to wait for the new N/C data to be read and decoded before entering the machine actuation registers. The hesitation that would be caused by slow transfer of data could cause the cutter to dwell, thereby marking the part surface.

RECALLING THE FACTS

1. With an N/C or CNC machine, the process starts with _____.
2. N/C data can enter the machine control through _____, floppy disks, _____, direct line to the CNC, or _____.
3. _____ are like storage mailboxes within the control that accept the N/C data consisting of machine coordinates and other codes, commands, and functions.
4. Buffer storage allows information to be stored in temporary holding _____ while an operation is being performed from the active machine registers.

UNIT 11

Drive Motors, Motion, and Feedback

KEY TERMS

drive motors *ballscrews* *closed-loop feedback systems*
open-loop *feedback systems* *resolution*

DRIVE MOTORS

In the previous unit we learned that after the N/C data is read, decoded, and sent to the machine actuation registers, some machine motion will occur. CNC machine motion is supplied by *drive motors* that move and position the moving components of the machine tool. Machine tool drive motors position the moving machine components the way the manual machine operator uses the hand wheels to position these components on manual equipment.

Machine tool drive motors are of four types: stepper motors, direct current (DC) servos, alternating current (AC) servos, and hydraulic servos. Stepper motors move a set amount of rotation (a step) every time the motor receives an electronic pulse. For example, a stepper motor may move 1/300 of a motor revolution with one pulse of electricity. The distance the axis moves is based on the number of electrical pulses the control generates. Motor speed is determined by how fast the pulses are sent to the motor by the MCU.

DC and AC servos are widely-used variable-speed motors found in modern CNC machines. Unlike a stepper motor, a servo does not move a set distance; when current is applied, the motor starts to turn, and when the current is removed, the motor stops turning. Servo motors allow complete control of:

- acceleration and deceleration
- distance rotated (based on input energy)

- motor speed
- reversing time

The AC is a fairly recent development. It can develop more power than a DC servo and is commonly found on CNC machining centers. The typical electric servomotor system has the feedback system and motor in the same housing, as seen in Figure 11–1. Recent advances in the power and response of electric motors have made them the preferred choice for all CNC machines.

Using electric motors permits a reduction in the size (or even total elimination of) the hydraulic system. Machines with electric drives may still use a small hydraulic system for tool clamping and changing.

Hydraulic servos, like AC or DC servos, are variable-speed motors. Because they are hydraulic motors, they are capable of producing much more power than an electric motor and have been used on large N/C machines in the past.

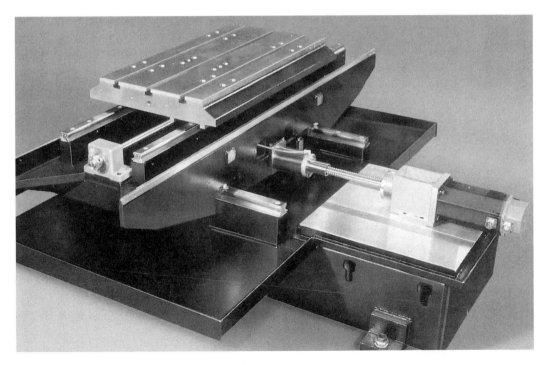

FIGURE 11–1
A CNC's electric servomotor. (Courtesy of Cincinnati Milacron, Inc.)

MOTION

The rotary motion from the drive motors is converted to linear (straight line) motion by recirculating ball-lead screws called *ballscrews.* This type of screw (Figure 11–2) has low friction and almost no backlash and can run at very high speeds. The drive motors and ballscrews work together to move the machine's components for the metal-cutting operations. The machine axes travel on low friction roller bearings to allow for higher positioning speeds. Older machines without roller bearings were typically limited to less than 100 in./min. maximum table speed.

Electrically driven machines can have table travel speeds in excess of 400 in./min. and as high as 1,200 in./min. or higher for certain types of CNC machines. These high table speeds are used for positioning the machine between machining operations, but actual machining seldom takes place at speeds above 150 in./min.

Resolution of a N/C or CNC machine is the smallest movement the machine is capable of making. Resolution is a combination func-

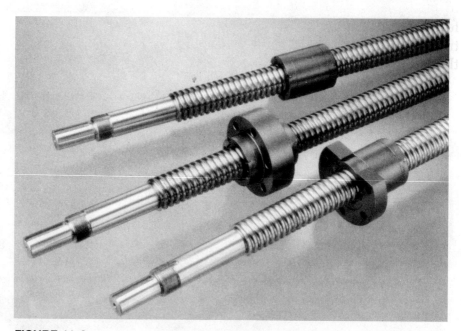

FIGURE 11–2
Recirculating ball-lead screws called ballscrews. (Courtesy of Thomson Industries, Inc.)

tion of the machine and control hardware. The control must be able to output very small movement commands while the drive motors are able to act upon the commands.

FEEDBACK

Once the command signals have been sent from the MCU to the machine, the slide motion and spindle movement occur. How then does the CNC know that the machine is properly positioned? Unless the control and machine form a *closed-loop feedback system*, the control really has no way of knowing if the machine is properly positioned. Such systems, as seen in Figure 11–3, are similar in operation to driving a car. When driving a car you will check the speedometer to determine speed. The actual speed is compared to the desired speed based on the speed limit signs. Your brain then determines the difference between the posted speed and the actual speed and your brain then tells your foot to ease up on the gas pedal until you reach the desired speed.

Open-loop feedback systems provide no check or measurement to indicate that a specific position has actually been achieved. No feedback information is passed from the machine back to the control. The main difference between open- and closed-loop systems is that with closed-loop systems, the actual output is measured and a

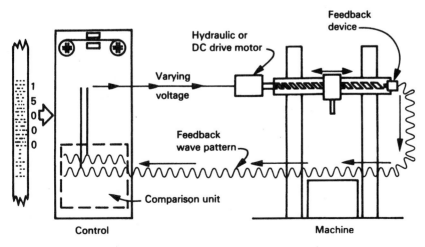

FIGURE 11–3
A closed-loop feedback system.

signal corresponding to this output is fed back and compared to the input machine registers. Virtually all CNC machines today are closed-loop systems.

RECALLING THE FACTS

1. CNC _____ supply motion to the moving components of the machine tool.
2. _____ feedback systems on CNC machine tools enable the control to tell that the machine tool is properly positioned.
3. _____ feedback systems provide no check or measurement to tell that the machine is properly positioned.
4. The main difference between _____-loop and _____-loop systems is that with closed-loop systems, the actual output is measured and a signal corresponding to this output is fed back and compared to the input machine _____.
5. _____ is the smallest movement the machine is capable of making.

UNIT 12

Axis Relationships and Control

KEY TERMS

axis	X-axis	quadrants
rectangular coordinates	C-axis	Z-axis
B-axis	Y-axis	A-axis

MACHINE/AXIS RELATIONSHIPS

So far in this section you have learned that:

1. N/C data must be input in some manner to the MCU,
2. the machine actuation registers accept the incoming positioning data,
3. the drive motors move the machine components to the data loaded in the machine registers and,
4. closed-loop feedback makes sure that the machine is positioned where the control thinks it is.

Now you need to learn how the positions that the machine moves to are determined. You need to know and understand this because knowing where the positions are and how to determine where the positions are is really what N/C programming is all about.

CNC machines position themselves along coordinate axes. *Axis* refers to any direction of motion that is totally controlled by specific N/C commands. The primary axis movements to be studied in terms of their relationships to each other at this time are the X- and Y-axes. Machines with only X and Y positioning capability are known as two-axis machines. An example of table movement in the X and Y direction is shown in Figure 12–1. If you were standing in front of this machine, *X-axis* movement would be the machine table moving from left to right and right to left. *Y-axis* movement would be the machine table moving in and out, toward you and away from

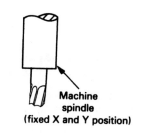
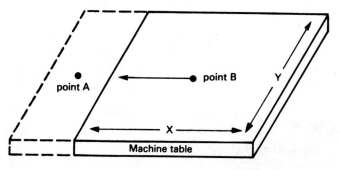

FIGURE 12–1
An example of table movement in the X and Y direction showing movement from point A to point B.

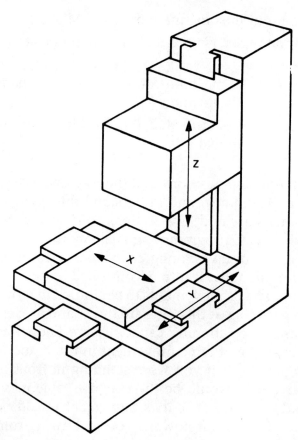

FIGURE 12–2
A vertical N/C machine showing X, Y, and Z axis directions.

you. On machines of this type, the machine spindle direction is up and down. This is called the **Z-axis**. Machines with Z-axis spindles that are up and down are called *vertical machines*. An example of the three vertical machine axis directions, X, Y, and Z, is shown in Figure 12–2. Z-axis motion is the movement of the spindle down into the workpiece or up away from it.

Horizontal CNC machines have a somewhat different set of directions for the X-, Y-, and Z-axes. If you were standing in front of a horizontal CNC machine, X-axis movement would be the machine table moving from left to right and right to left. Y-axis movement would be the spindle carrier or machine table moving up and down. Z-axis movement would be the spindle moving in and out, toward you and away from you. Machines with Z-axis spindles that move in and out are called *horizontal machines*, and have somewhat different axis directions of X, Y and Z, as shown in Figure 12–3. Z-axis direction changes due to the fact that CNC machines are made with vertical and horizontal spindles.

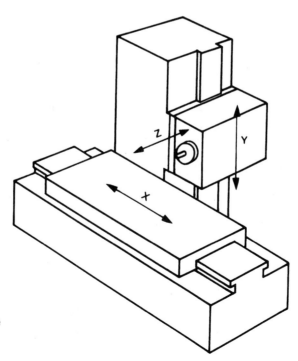

FIGURE 12–3
A horizontal N/C machine showing *X, Y,* and *Z* axis directions.

To help understand the Z-axis, it can be said that a line through the center of the machine spindle is actually the Z-axis. It is only when the actual depth of cut (Z-axis) is controlled through programming that the machine is considered a true three-axis CNC machine. This means that the machine is capable of motion in X, Y, and Z directions at the same time. Additional machine/axis relationships can be seen in Figures 12–4 and 12–5.

As the CNC machine positions itself corresponding to the programmed positions input to the MCU, the positions obtained and in the machine actuation registers are displayed on the CNC's display screen (the cathode ray tube or CRT screen). An example of a CNC's CRT screen is shown in Figure 12–6. As you can see, other information is displayed on the CRT screen that will be discussed in detail later.

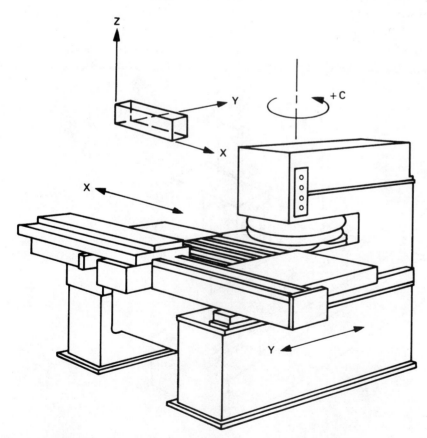

FIGURE 12–4
An N/C turret punch press.

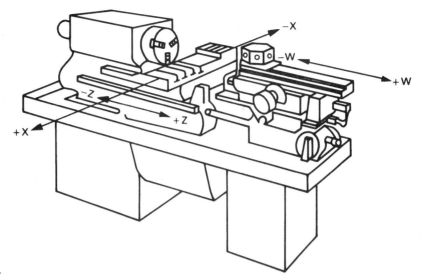

FIGURE 12–5
An N/C turret lathe.

FIGURE 12–6
A typical CNC unit with CRT screen displaying active control/machine information. (Courtesy of Cincinnati Milacron, Inc.)

113

POSITIONING IN QUADRANTS

This entire concept of numerical control is based on the principle of *rectangular coordinates* discovered by the French philosopher and mathematician, Rene Descartes. This mathematical development made over 300 years ago is better known as the *Cartesian coordinate system*. Through the use of rectangular coordinates, any specific point in space can be described in mathematical terms along the X-, Y-, and Z-axes.

The system of rectangular coordinates is illustrated in Figure 12–7. The X-axis is horizontal (left and right) and the Y-axis is vertical (up

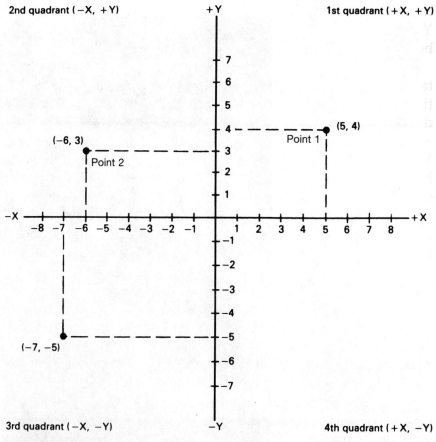

FIGURE 12–7
Cartesian coordinate system.

and down on the page). The Z-axis is then applied by holding a pencil over this example with its point at the location where the X and Y lines cross each other. The point where the X and Y lines cross each other is called the *origin*, or *zero point*. Four **quadrants** are formed when the X- and Y-axes cross. Each quadrant is numbered in counterclockwise rotation, as seen in Figure 12–7.

The plus (+) and minus (−) signs indicate direction from the zero point along the X- and Y-axes. A move to the right along the X-axis is a move in the positive X direction. The value of X increases as you move to the right. A move to the left along the X-axis is a move in the negative X direction. The value of X decreases as you move to the left. A move up along the Y-axis is a move in the positive Y direction. The value of Y increases as you move up toward the top of the page. A move down along the Y-axis is a move in the negative Y direction. The value of Y decreases as you move down toward the bottom of the page.

To get to point 1 in Figure 12–7 and describe it in terms of its rectangular coordinates, you count over +5 places to the right along the X-axis, and up +4 places along the Y-axis. To get to point 2, and describe it in terms of its rectangular coordinates, you count over −6 places to the left along the X-axis and up +3 places along the Y-axis. As seen from Figure 12–7, all positions in the first quadrant have positive X and positive Y values (+X, +Y). All positions in the second quadrant have negative X and positive Y values (−X, −Y). Points in the third quadrant will have values of negative X and negative Y (−X, −Y). In the fourth quadrant, all positions will have values of positive X and negative Y (+X, −Y).

In numerical control programming, plus signs may or may not be printed if the value is positive. A plus sign normally does not need to be printed if the value is positive unless it is required by that particular CNC machine and control type. This is because the CNC control assumes the value is positive unless a minus sign is in front of the value. Minus signs must be written to distinguish between negative and positive values.

Figure 12–7 illustrates two important concepts. The first is how points are plotted along the X- and Y-axes and that both an X and a Y value are needed to describe one point. The second is that a plus or minus sign determines the quadrant in which the point will fall.

Both concepts are important because:

1. X and Y values are the positions that the CNC machine moves to in a metal-cutting operation. Wrong X and Y values will cut the part wrong and perhaps wreck the machine, damage the part and cutter, and possibly cause injury.
2. CNC machines today can be programmed in any quadrant. Because of this, whether a value is plus or minus is very important because even though an X or Y value is correct, if the sign (+ or −) is incorrect, the net effect is a wrong position and a wrecked machine, damaged part, etc. What quadrant the machine will be programmed in is a function of how the machine axes and part are aligned. This will be discussed later.

At first glance, it appears that it would be easier if all work could be done in the first quadrant since all values are positive and no negative signs would be needed. However, any of the four quadrants may be used. Therefore, you must be familiar with the use of both plus and minus signs in all four quadrants.

MACHINE CONTROL IN THE COORDINATE SYSTEM

Now you need to look closer at two-axis control on a vertical N/C machine. Two-axis control on vertical N/C machines normally consists of X- and Y-axes. Since most of the vertical N/C machines today are three axes (X, Y, and Z), you need to first look at an older but simpler two-axis N/C machine, as shown in Figure 12-8. In most cases, as you learned earlier, the machine table moves left and right to position the cutter in the X direction. The machine table moves in and out on a saddle for the Y-axis direction. In order to obtain identical X and Y movements on some two-axis machines, the table remains stationary and the spindle moves to satisfy X and Y locations. The movement of the cutter remains the same regardless of whether the spindle or table positions the workpiece. This consideration is taken care of by the machine and control manufacturers. The programmer must specify only the locations and the plus or minus signs in relation to the zero point.

Z-axis movements, just like X and Y movements, are programmable. As explained earlier, a line through the center of the machine

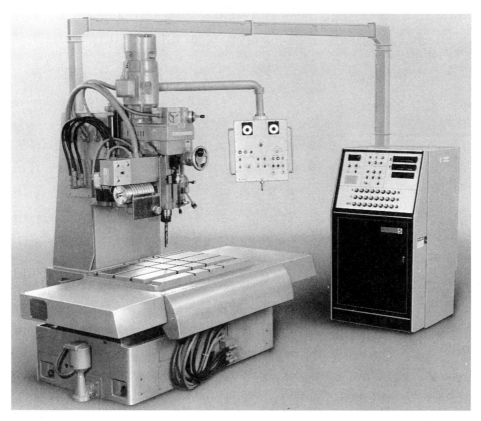

FIGURE 12–8
A typical two-axis N/C machine. (Courtesy of Cincinnati Milacron, Inc.)

spindle is the Z-axis. Figure 12–9 illustrates a vertical Z-axis with the X- and Y-axes drawn in relation to the workpiece and machine table. A positive Z movement moves the tool away from the work. A negative Z movement moves the tool into the work. In this example, (0, 0) is the left front corner of the machine table as you would face the machine. If you think of (0, 0) in relationship to the four quadrants, this means that all positions for this part will be in the first quadrant and have positive X and positive Y values. Figure 12–10 illustrates the same first quadrant example only with a horizontal Z-axis drawn in relation to the workpiece and machine table.

Some of the more advanced CNC machines are equipped with four- and five-axis contouring capabilities. These capabilities gener-

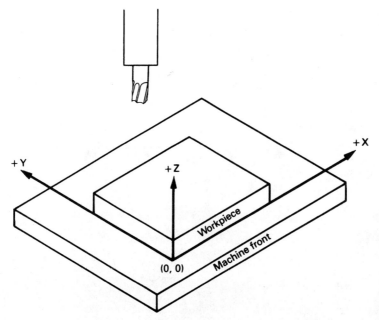

FIGURE 12–9
Vertical Z-axis drawn in relation to the X and Y axes, workpiece, and machine table.

ally involve second and third axes that run in the same direction as the X-, Y-, and Z-axes plus some additional movements. Rotational movements consist of three primary axes of motion: A, B, and C. Rotary axes of motion rotate around the three primary axes of motion, X, Y, and Z, as seen in Figure 12–11. The *A-axis* rotates around

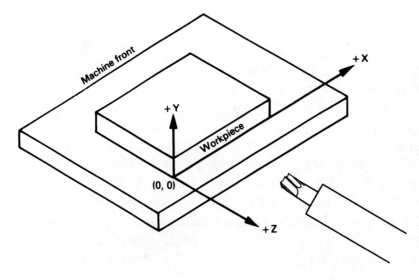

FIGURE 12–10
Horizontal Z axis drawn in relation to the X and Y axes, workpiece, and machine table.

AXIS RELATIONSHIPS AND CONTROL • 119

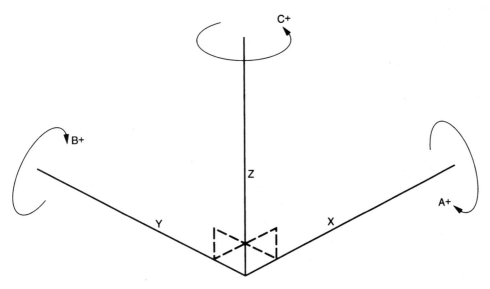

FIGURE 12–11
Rotary axis of motion: A-axis rotates around a line parallel to the X axis, B around Y, and C around Z.

a line parallel to the X-axis. The **B-axis** rotates around the Y-axis, and the **C-axis** rotates around the Z-axis.

RECALLING THE FACTS

1. _____ refers to any direction of motion that is totally controlled by specific N/C commands.
2. If you were standing in front of a vertical drilling machine, the _____ would be the movement of the machine table left and right, and _____ would be the movement of the machine table in and out.
3. A line through the center of the machine axis is always the _____.
4. The entire concept of numerical control is based on the system of _____ coordinates.
5. CNC machines today can be programmed in any _____.

6. The _____-axis rotates around a line parallel to the X-axis; the B-axis rotates around a line parallel to the _____-axis; and the _____-axis rotates around a line parallel to the Z-axis.

7. List what quadrant (1, 2, 3, or 4) the following points are in:
   a. X + 5.3, Y − 2.5 _____
   b. X + 20, Y + 12.2 _____
   c. X − 6.8, Y + 1.5 _____
   d. X − 8.4, Y − 5.7 _____
   e. X + 2.9, Y − 3.3 _____
   f. X − 4.3, Y + .5 _____

8. A _____ Z moves the tool away from the work, and a _____ Z moves the tool into the work.

9. Fill in the coordinate values for the following points.
   a.

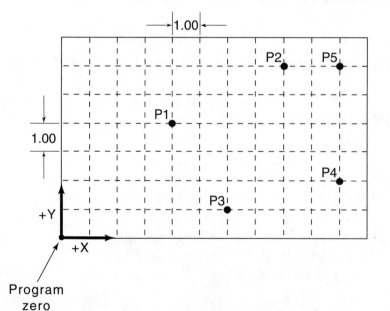

Point	X	Y
P1	4.00	4.00
P2		
P3		
P4		
P5		

AXIS RELATIONSHIPS AND CONTROL • 121

b.

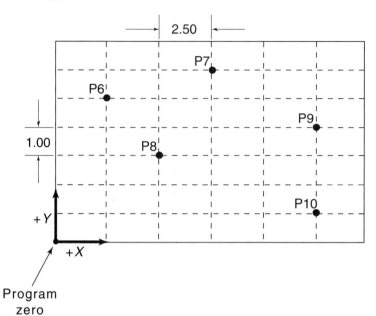

Point	X	Y
P6	2.50	5.00
P7		
P8		
P9		
P10		

c.

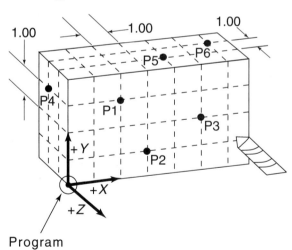

Point	X	Y	Z
P1	2.00	3.00	0
P2			
P3			
P4			
P5			
P6			

d.

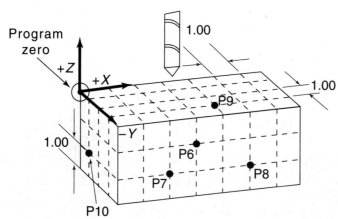

Point	X	Y	Z
P6	3.00	−4.00	−1.00
P7			
P8			
P9			
P10			

UNIT 13

Types of CNC Systems

KEY TERMS

incremental dimensioning   full floating zero   absolute dimensioning
incremental programming   delta   absolute programming

INCREMENTAL VS. ABSOLUTE SYSTEMS

When studying a part print in order to program a part, one of the first things a programmer will notice is how the part is dimensioned. Carefully study the part in Figure 13–1. The distance from the left edge of the part to hole 1 is 1.25. From hole 1 to hole 2, the distance is 1.50. The distance from hole 2 to hole 3 is 1.50, and 1.62 from hole 3 to hole 4. This is known as *incremental dimensioning*. It is also referred to as *delta* dimensioning. The word *delta* is derived from a Greek letter that means the difference between two quantities. In Figure 13–1, each dimension is given incrementally from the last position to the next position.

Incremental programming works according to the same principle; it positions the work or cutter from the point where it currently is to the next point programmed. Calculations are made from the tool's present location to the next position where it is going. The use of plus or minus signs takes on a new meaning when used in the in-

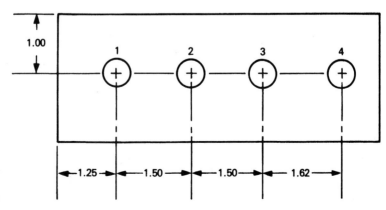

FIGURE 13–1
Incremental dimensioning.

cremental mode. A positive X move does not refer to a specific rectangular quadrant, but directs the tool to move to the right along the X-axis from its current position. A negative X move directs the tool to the left. Similarly, a positive Y move positions the cutter up from the present location, and a negative Y is a command to move down. A positive Z directs the cutter away from the workpiece, while a negative Z is a move toward or into the workpiece.

Looking closely at the workpiece in Figure 13–2 you will see how closely it resembles the one shown in Figure 13–1. The difference is the way the actual part is dimensioned. This part is dimensioned in a manner known as *absolute* or *baseline dimensioning* because all positions are given as distances from the same zero location or reference point. All dimensions are calculated from one zero point, as indicated in Figure 13–2.

Absolute programming operates similarly to absolute dimensioning. All positions are figured from the same zero or reference point. All positional moves come from the same reference or "datum edge" at all times, as opposed to an incremental system, where each new move is an incremental distance from the last.

One advantage of absolute systems over incremental systems concerns positioning errors. If a positioning error occurs in an incremental system, all the rest of the positions are affected and all remaining moves will be wrong. This is because each new position is an incremental move from the last. When a positioning error occurs

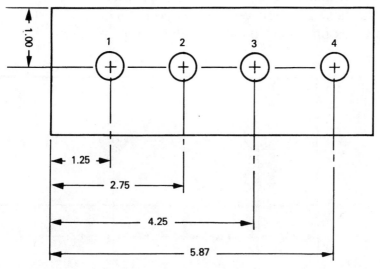

FIGURE 13–2
Absolute dimensioning.

in an absolute N/C system, one particular position or location will be wrong but remaining positions are not affected. This is because all dimensions and each remaining positional move are always based from the same zero or reference point. This is not to say, however, that absolute dimensioning and programming is better than incremental dimensioning or programming. This philosophy started in the early days of N/C when purchasers were forced to choose between either an absolute or an incremental system. This resulted in part drawings having to be dimensioned (absolute or incremental) to suit the type of N/C system being used to machine the part. If the part drawing was not or would not be dimensioned to match the N/C system being used to machine the part, the programmer was forced to do all the math and convert dimensions from one system to another. This process was very error-prone.

Both absolute and incremental systems have their logical areas of application and neither is always right or wrong. There are certain applications in which both systems can be used most efficiently, sometimes even within the same program. CNC controls today are capable of working in either mode with just a simple instructional code inserted to make the change. With the capability of modern controls, the controversy over which is better is of little importance. In most cases, it is up to the programmer to decide which system to use for the part being programmed. However, programmers must have a thorough understanding of both modes and be able to make the best use of each.

BEGINNING FROM ZERO

Before starting to program, the programmer must establish where the zero position will be set. The zero position can be set anywhere on CNC machines today as long as the machining positions stay within the X, Y, and Z range of the machine tool. It is generally up to the programmer to decide where the zero point or origin will be set in any part program. In many cases, the zero point has to be "shifted" because the zero positions are not the same from one part program to any other.

Most CNC machine tools today are equipped with *full floating zero*. Full floating zero allows the operator to locate the workpiece in any convenient location on the machine table. Once the work-

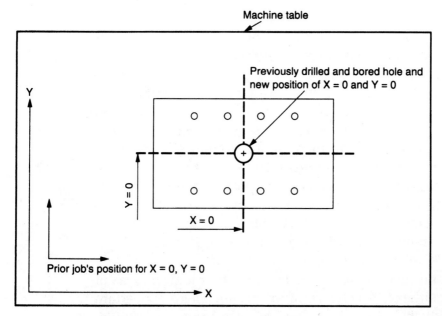

FIGURE 13-3
Full floating zero CNC system.

piece is set up or positioned on the machine table, the operator then obtains the alignment positions for the particular part program from the program manuscript. In Figure 13–3, the alignment positions in X and Y are X = 0 and Y = 0 (0,0). These values are then dialed into the control. The operator depresses the cycle start button, and the machine rapid traverses to some location which, at this point, may not be relative to the workpiece setup just completed. The operator then adjusts the zero location of the machine table to the setup location just completed. This moves the machine axes without changing the actual readout on the control for X and Y. Zero shifting the machine merely operates the table drive motors in X and Y, but their signal does not enter the memory section of the control unit. Adjusting zero shift for this particular workpiece moves the table zero to the tram position.

Once the new workpiece is "zeroed" or "trammed" in, the machine slides are aligned and locked to the new alignment positions. The operator can then consistently run parts according to N/C data commands relative to the manuscript alignment position and the convenient zero location.

Full floating zero greatly enhances actual machine spindle cutting time by reducing setup time. The programmer and operator gain flexibility because the programmer now can make zero any place on the machine table, enabling positive and negative programming.

RECALLING THE FACTS

1. In incremental dimensioning, also known as _____ dimensioning, each dimension is given incrementally from the last position to the next position.
2. In _____ dimensioning, all positions are given as distances from the same zero or reference point.
3. In incremental programming, each position is figured from the point where the cutter currently _____ to the _____ position programmed.
4. In _____ programming, all cutter positions are from the same zero or _____ point.
5. Full floating _____ allows the operator to locate the workpiece in any convenient location on the machine table. Once the workpiece is set up or positioned on the machine table, the operator then obtains the alignment positions for the part program from the program _____.
6. Practice with absolute locations.

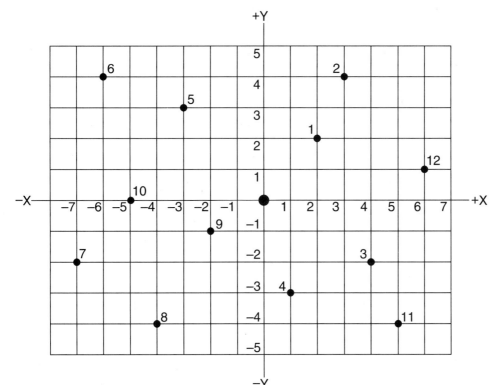

Point	X Position	Y Position
1	2.00	2.00
2		
3		
4		
5		
6		
7		
8		
9		
10		
11		
12		

7. Practice with incremental locations.

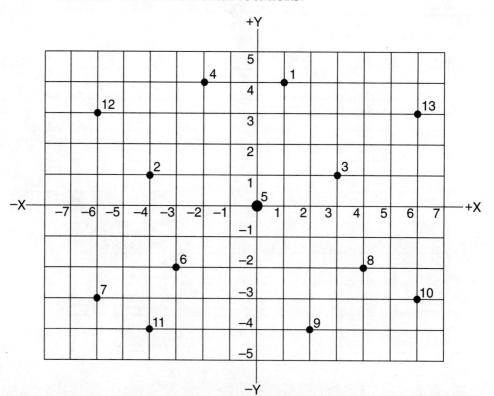

Points	X Distance	Y Distance
5 TO 3	3.00	1.00
13 TO 1		
12 TO 9		
10 TO 8		
6 TO 13		
2 TO 4		
9 TO 3		
1 TO 4		
7 TO 13		
6 TO 5		
2 TO 10		
4 TO 9		

UNIT 14

N/C Data Input and Storage

KEY TERMS

punched tape	*cassette tape*	*floppy disk*
DNC	*manual data input (MDI)*	

A variety of control systems have been developed since the beginning of numerical control. Consequently, a wide variety of different methods for N/C data input has also been developed. In order to standardize this N/C information, an Electronics Industries Association (EIA) subcommittee was formed many years ago to recommend a set of standards which would be acceptable to control system manufacturers and machine tool manufacturers and users.

Several types of input media were tried as numerical control evolved. Initially, the most common types were punched cards, magnetic tape, and punched tape. Punched cards and magnetic tape (on open reels) proved highly impractical for shop use. Eventually, industry adopted the punched tape as standard because it was more suitable for shop environments, and adopted a standard coded tape format which will be discussed in the next unit.

Today, N/C data may be passed to the MCU in five different ways:

1. punched tape
2. cassette tape
3. floppy disk
4. manual data input (MDI)
5. DNC

PUNCHED TAPE

Tape readers (Figures 14–1 and 14–2) utilize punched tape input and are classified as mechanical or photoelectric (light) readers. They

FIGURE 14–1
A typical CNC cabinet with a tape reader and tape reels. (Courtesy of Cincinnati Milacron, Inc.)

may read a single row of information at a time (each line of holes across the tape represents one character), or they may read a complete *block* of instruction (several characters). Reader speeds will vary considerably. Older mechanical readers fed the tape through the reader by means of sprockets and sprocket holes in the tape. These were capable of reading approximately sixty characters per second. Photoelectric readers are used today and can read approximately 300 to 500 characters per second.

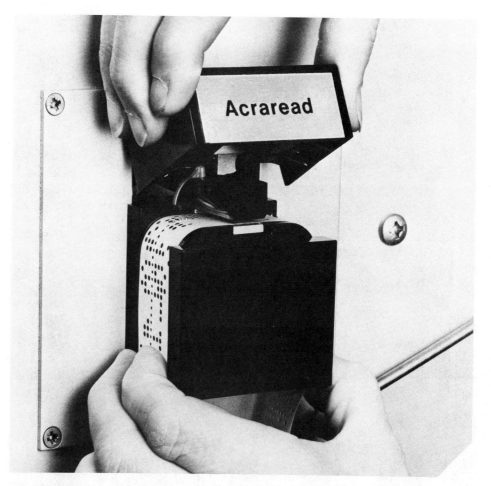

FIGURE 14–2
Loading a tape reader. (Courtesy of Cincinnati Milacron, Inc.)

Photoelectric readers are the most commonly used because of their speed. Photoelectric readers operate by light beams which pass through the holes in the tape and show up on a photocell. The light beams are then converted to electrical impulses and are passed on to the controller providing smooth and continuous motion of the machine tool.

Punched tape has appeared in several different types of materials, sizes, and coding systems. The tape materials are primarily grouped under three main headings: paper, Mylar (Du Pont

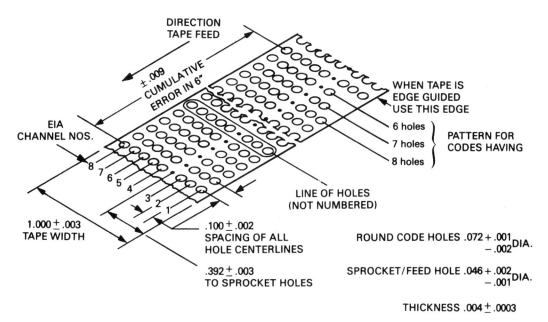

FIGURE 14–3
Standard 1 in. wide, eight track tape with dimensions and tolerances shown.

Company's trade name for a tough plastic), and foil. There are, however, other combinations and variations of these three.

The specifications for the size of punched tape have varied tremendously since the early years of N/C development. Standardization became necessary in order to cut costs and establish common methods of programming among machine tool manufacturers. This was done through the cooperation of the EIA and the Aerospace Industries Association (AIA). Standardization covers two important categories: character coding and physical dimensions.

As shown in Figure 14–3, the physical dimensions of the tape have been standardized—one inch in width with eight tracks. It was determined that these tracks, or *channels*, were to run the length of the tape. The actual dimensions for thickness, hole spacing and size, and tolerances were also established at that time. Figure 14–4 shows all character codes and the appropriate punches for one-inch wide, eight-track tape (EIA-RS-244).

In the early days of N/C, actual tape preparation (punching) was done by manual tape preparation equipment similar to typewriters

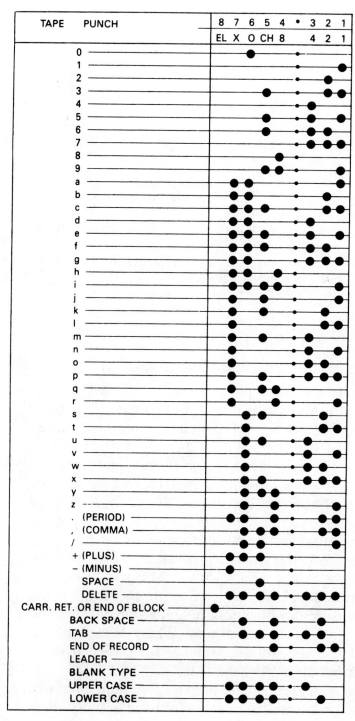

FIGURE 14–4 Character codes and punches for the EIA/BCD system (EIA RS-244).

FIGURE 14–5
Computer tape-punching equipment. (Courtesy of Flexmate, Inc.)

called "Flexowriters." Today, N/C tape punching, although still used by some manufacturers, is done by means of computer tape-punching equipment, as seen in Figure 14–5.

CASSETTE TAPES AND FLOPPY DISKS

Cassette tapes, similar to those bought in music stores and used in automobiles and stereo equipment (Figure 14–6), are used today by some manufacturers for inputting and storing N/C part program data.

Additionally, floppy disks, similar to those used on personal computers (Figure 14–7), are also a preferred choice of input and storage media for CNCs. Normal storage capacity ranges from 100–200 feet of part program storage, but some double-density disks, for example, can hold considerably more. However, two common problems still exist for cassette tapes, floppy disks, and punched tape—storage and revision control. Cassette tapes, floppy disks, and punched tape must be stored in bins or file cabinets and reused when

FIGURE 14–6
A typical cassette used for loading and storing CNC data.

required. This introduces several problems. They can be lost, misplaced, misfiled, or damaged, and they must be kept up-to-date with the latest part print and planning changes. Regardless of the type of external storage medium used, trying to store, manage, and keep up-to-date with the latest revision of part program changes is an administrative nightmare. Failure to accurately maintain the current

FIGURE 14–7
A floppy disk used for loading and storing CNC data.

FIGURE 14-8
Manual data input (MDI). (Courtesy of Okuma America Corporation.)

revision level for N/C data could cause an outdated version of the part program to be run, thereby possibly scrapping parts, causing tool, machine, or part damage, and possible operator injury.

MANUAL DATA INPUT

Manual data input (MDI) is a means of programming the machine tool through the CNC keyboard and push-buttons. MDI input, as seen in Figure 14-8, can range from just positioning the X-, Y-, and Z-axes to inputting a complete part program and running that program from memory. MDI is completely operator-controlled and, therefore, is subject to human error. Speed and accuracy depend entirely on the operator's ability to find, interpret, and input the correct information. All CNC machines today are capable of accepting and storing manual part programs.

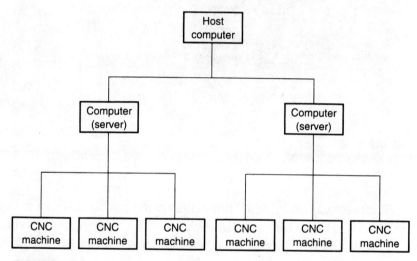

FIGURE 14-9
A comparison of direct and distributed numerical control.

DNC

Many CNCs in use today are wired directly to a central or personal computer for direct downloading of N/C data to the MCU. N/C programs can be generated, stored, and edited on the computer and then downloaded to each machine as needed. Such systems are called *DNC* for Direct or Distributed Numerical Control. A comparison of the differences between direct and distributed numerical control is illustrated in Figure 14-9.

DNC systems bypass any punched tape, cassette, or floppy disk reader and use a standard RS-232-C communications port and coaxial cable for connecting the computer to the MCU. Older N/C and CNC controls do not have the RS-232-C communications port; they require a direct connection in place of the tape reader called a *BTRI* (Behind the Tape Reader Interface). Direct numerical control systems only download a small portion of the program each time a part is run, feeding the machine with N/C data much like a tape reader. Distributed N/C systems download the entire program into the MCU's memory.

Typically, the CNC has an internal memory for program storage. So, depending on the MCU and its input media, it is possible to "read in" and store part program data from punched tape, cassette tape, floppy disk, MDI, or DNC (distributed N/C) and run the program continuously from memory. An example is illustrated in Figure 14–10. Storing 200 feet of punched tape is common for many CNCs but some modern controls will take up to 2,000 feet.

Primary advantages of DNC (distributed N/C) systems include:

> **"Tech Talk"**
>
> When a CNC control is tested, a key measurement is the time between any electronic failure called *mean time between failure*, or MTBF. In 1988, the MTBF for a typical hard drive was 20,000 hours. Today that number has risen to 200,000 hours. This means that, on average, a new CNC hard drive, running continuously, could be expected to have one failure every 22.8 years.

1. elimination of punched tape, floppy disks, and cassette tapes as a medium for storing and loading programs,
2. easier to keep track of and control program revisions and tool lists,
3. centralized and on-line location of part programs for direct access and retrieval,
4. improved program security,
5. on-machine changes to the part program can be uploaded to the central computer after the part is run, and
6. the DNC system can be used for other plant-wide data collection and distribution.

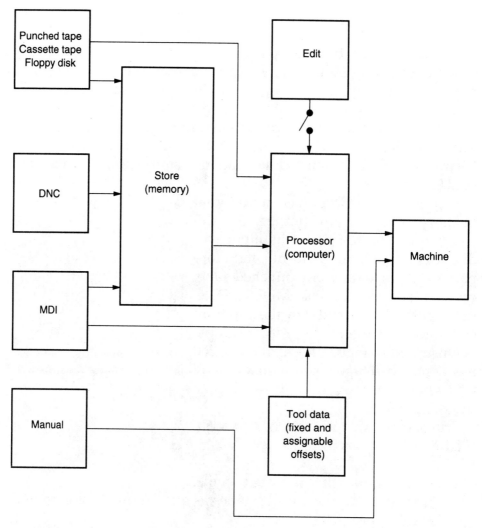

FIGURE 14–10
How a CNC unit functions with its microcomputer.

RECALLING THE FACTS

1. In order to use _____ input, a CNC control must have a _____ reader. These may be either mechanical or _____, with _____ being the most widely used on CNCs today.
2. _____ tapes and _____ disks are two other types of media used today to store part programs.

3. The two common problems for _____ tapes and _____ disks are _____ and _____ control.
4. Failure to maintain the proper _____ level for N/C data could cause an outdated version of the part program to be run.
5. _____ is a means of programming the machine tool through the CNC keyboard and push-buttons.
6. DNC stands for _____ _____ _____.
7. With DNC systems, a direct line is run from the CNC unit to the storage _____ and any punched tape, cassette tapes, or floppy disks are bypassed.

UNIT 15

N/C and CNC Formats, Codes, and Downtime

KEY TERMS

tape format
word address format
American standard code for information interchange (ASCII)

heat
vibration
power disturbances
oxidation

decimal point format
binary-coded decimal (BCD)
contaminated air

TAPE FORMATS

The *tape format* is the general sequence and arrangement of coded information on a punched tape. This information conforms to EIA standards and appears as words made of individual codes written in horizontal lines, as shown in Figure 15-1. For example, there are five words that make up a block (one instruction) for this particular tape format. The most common type of tape format in current use is the *decimal point format*. In the decimal point format, word addresses are used for each word but decimal points are programmed. An earlier format still in use today without the decimal point is the word address interchangeable or compatible format. Fixed sequential, tab ignore, and tab sequential are older formats and are obsolete today. Examples of these tape formats are illustrated in Figure 15-2.

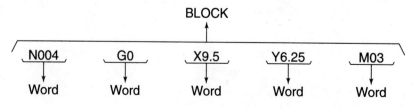

FIGURE 15-1
A block of information and the individual words.

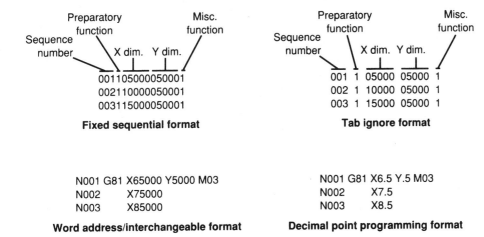

FIGURE 15-2
Examples of tape formats.

The *word address format*, standardized by the EIA years ago, uses a letter address to identify each separate word. This is similar to each house having a particular address in order for the post office to deliver mail to the right houses. The difference is that houses have a numerical address and N/C words have a letter address. By assigning an alphabetical address to each coordinate and function word, the block format becomes more flexible. That is, the words can appear in the block of instruction in any order because each word has its address which corresponds to its own register location (as you learned earlier) in the control. As a result, words do not have to appear in a rigid sequence (Y following X, Z following Y, and so forth), and tapes are more interchangeable with machines in the same class.

The letter address for N/C words has two purposes:

1. identify the N/C word with its matching register location in the control and,
2. minimize the amount of N/C data storage. This means that as new N/C data is read and changed, codes that change from one block to another are the only ones that need to be reprogrammed; repeated codes can be omitted. For example, if X + 3.750 and Y + 1.125 are programmed and the Y position

changes to Y + 2.250, only the Y value of Y + 2.250 needs to be programmed on the next line. X + 3.750 remains in the X register until a new X value enters the control and cancels out the current value of X + 3.750

Word addresses typically conform to the ANSI/EIA RS-274-D standard, as seen in Figure 15–3. This standard is followed by most CNC equipment manufacturers.

It should be pointed out that even though the general tape format is followed, the specific tape format for a particular machine tool and its specific control type must also be followed. For example, a CNC lathe and a CNC machining center are two different and unique CNC machines. Both can be programmed using the decimal point programming format, but the CNC lathe tape input will not work in the machining center and vice versa. This is because each machine and control system has its own set of words and registers that are recognizable. If the CNC control is fed words which it cannot recognize, e.g., an N/C lathe tape accidentally loaded in a machining center control, a system failure will occur and the machine will not make any movements. Therefore, it is always important to understand the specific tape format for the particular machine and control type being programmed.

Standard Word Addresses

A	Rotation about the X-axis
B	Rotation about the Y-axis
C	Rotation about the Z-axis
F	Feed rate commands
G	Preparatory functions
I	Circular interpolation X-axis offset
J	Circular interpolation Y-axis offset
K	Circular interpolation Z-axis offset
M	Miscellaneous commands
N	Sequence number
O	Sequence number for secondary axis commands
R	Arc radius
S	Spindle speed
T	Tool number
X	X-axis data
Y	Y-axis data
Z	Z-axis data

FIGURE 15–3
Standard word address followed by most CNC equipment manufacturers.

CODING SYSTEMS

The Electronics Industries Association (EIA) developed the *binary-coded decimal (BCD)* system for coding tapes several years ago. This is a binary (which means "two") system of number representation in which each decimal digit is represented by a group of binary digits forming a character. It incorporates the best features of both the binary and decimal systems and gives a compact, easily simulated method of converting decimal dimensions into the analog voltage ratios which control the machine tool.

American Standard Code for Information Interchange (ASCII) was another coding system developed and is now the standard coding system used throughout the computer and communication industries. This particular code was compiled by a committee from several different groups working with the United States of America Standards Institute. The overall objective of this group, now named American National Standards Institute (ANSI), was to obtain one coding system which will be an international standard for all information processing and communication systems.

There are several coding differences between BCD (EIA) and ASCII (pronounced "AS-key"). However, modern CNCs can automatically detect and adjust for either EIA or ASCII input. ASCII provides coding for both uppercase and lowercase letters, while BCD codes are the same for both. The ten-digit codes (0 through 9) in ASCII are the same as BCD coding, but holes are punched in two additional tracks to identify the numbers and certain symbols. The ASCII letter codes, however, are quite different from those used in BCD. A comparison of the two systems is shown in Figure 15–4.

At this point it is only important for you to know that there were two coding systems at one time and that ASCII is the international standard for all CNC, information processing, and communication systems. Although it is important to generally see and understand what the codes are and how they were developed, do not worry about knowing or memorizing the hole locations, codes, or values for each letter and number in the example.

DOWNTIME

As with any electronic or mechanical equipment, problems occasionally occur and equipment malfunctions or does not work at all.

BCD (RS-244)									CHARACTER	ASCII (RS-358)								
		●			•				0			●	●		•			
					•			●	1	●		●	●		•			●
					•		●		2	●		●	●		•		●	
			●		•		●	●	3			●	●		•		●	●
					•	●			4	●		●	●		•	●		
			●		•	●		●	5			●	●		•	●		●
			●		•	●	●		6			●	●		•	●	●	
					•	●	●	●	7	●		●	●		•	●	●	●
				●	•				8	●		●	●	●	•			
			●	●	•			●	9			●	●	●	•			●
	●	●			•			●	a		●				•			●
	●	●			•		●		b		●				•		●	
	●	●	●		•		●	●	c	●	●				•		●	●
	●	●			•	●			d		●				•	●		
	●	●	●		•	●		●	e	●	●				•	●		●
	●	●	●		•	●	●		f	●	●				•	●	●	
	●	●			•	●	●	●	g		●				•	●	●	●
	●	●		●	•				h		●			●	•			
	●	●	●	●	•			●	i	●	●			●	•			●
	●		●		•			●	j	●	●			●	•		●	
	●		●		•		●		k		●			●	•		●	●
	●				•		●	●	l	●	●			●	•	●		
	●		●		•	●			m		●			●	•	●		●
	●				•	●		●	n		●			●	•	●	●	
	●				•	●	●		o	●	●			●	•	●	●	●
	●		●		•	●	●	●	p		●		●		•			
	●		●	●	•				q	●	●		●		•			●
	●			●	•			●	r	●	●		●		•		●	
		●	●		•		●		s		●		●		•		●	●
		●			•		●	●	t	●	●		●		•	●		
		●	●		•	●			u		●		●		•	●		●
		●			•	●		●	v		●		●		•	●	●	
		●			•	●	●		w	●	●		●		•	●	●	●
		●	●		•	●	●	●	x	●	●		●	●	•			
		●	●	●	•				y		●		●	●	•			●
		●		●	•			●	z		●		●	●	•		●	
	●	●		●	•		●	●	.			●		●	•	●	●	
		●	●	●	•		●	●	,	●		●		●	•	●		
		●	●		•			●	/	●		●		●	•	●	●	●
	●	●	●		•				+			●		●	•		●	●
	●				•				–			●		●	•	●		●
			●		•				Space	●		●			•			
	●	●	●	●	•	●	●	●	Delete	●	●	●	●	●	•	●	●	●
●					•				Car. Ret. or End of Block	●				●	•	●		●
		●		●	•		●		Back Space			Not Assigned						
		●	●	●	•	●	●		Tab					●	•			●
				●	•		●	●	End of Record			Not Assigned						
					•				Leader			Not Assigned						
	●	●	●	●	•	●			Upper Case			Not Assigned						
	●	●	●	●	•		●		Lower Case			Not Assigned						

BCD - Binary Coded Decimal system
ISO - International Standards Organization
ASCII - American Standard Code for Information Interchange
The systems ISO and ASCII are the opposite of BCD in that they are even-parity coded.

FIGURE 15–4
A comparison of EIA/BCD and ASCII coding systems.

When an N/C or CNC machine is not working or malfunctioning for some reason, it is referred to as being *down*. Referring to the time that an N/C or CNC machine is down is known as *downtime*. The primary causes of N/C, CNC, and other computer equipment downtime are:

1. heat,
2. power disturbances,
3. contaminated air,
4. vibration, and
5. oxidation.

The MCU and associated machine tool downtime these factors cause, in many cases, can easily be prevented if proper planning and prevention methods are followed. Case histories clearly indicate that controlling the environment within which N/Cs and CNCs operate is the most cost-effective way to increase uptime and reduce maintenance.

Heat

Electronic controllers and computer systems use a variety of integrated circuits (ICs), transistors, diodes, and other components—each with its own heat-tolerance level. Each component operates between a specific low/high temperature limit, above or below which they fail. Therefore, MCU cabinets housing these components must maintain a temperature that is well within each component's high and low cutoff limit. In addition, temperature should not fluctuate too much since fluctuations are responsible for servodrive misalignments and can cause condensation and oxidation.

MCU cabinet air conditioners should maintain a specific temperature between 75 and 100 degrees Fahrenheit (°F). When shop air raises cabinet temperature above 100°F, the air conditioner should step up cooling. Temperature gauges on MCU cabinets can provide maintenance personnel an accurate measure of conditions inside each cabinet.

Power Disturbances

Electrical supply power disturbances, usually in the form of electrical current power surges and spikes, are a major source of electronic and computer equipment failure and need to be eliminated through currently available suppression devices. Suppressors lessen transient electrical energy once it exceeds the nominal peak voltage of the electronic equipment.

Contaminated Air

Many electronic control cabinets are insufficiently sealed against shop air. Layers of oily dirt or black iron filings accumulate on circuit boards and often cause permanent hardware damage. The dust enters cabinets through wire ducts, vent panels, around switches, and door edges if the cabinet is improperly sealed.

Shop air circulated through cabinets is a major cause of dirt accumulation inside cabinets. To keep dirt out of the cabinet, all ventilation systems must be sealed off, and the air conditioner must be required to both cool and recirculate air. Doors and edges should be sealed off well with commercially available foam-rubber strips, and cable holes should be sealed with nonflammable sealing material.

Vibration

In addition to dirt, physical vibration continues to be a major reason for controller and computer failure. Circuit boards can gradually vibrate out of their sockets causing intermittent or lost connections and, consequently, equipment breakdowns. Vibration also shears components, as well as lead and plug socket connections over time. Shock-absorbing pads under cabinets can help reduce vibration.

Oxidation

Nonfunctional electrical/electronic connections can also be caused by normal aging or oxidation of a component. Some solutions are available to clean oxidized connectors but may cause them to corrode faster or form a thin film, thereby attracting dirt.

Computer and electronic equipment work best in the proper environment. In a business-computer installation, floors are often raised and insulated, and a lot of effort goes into planning the right air and power utilization levels. However, shop applications typically do not receive the same level of consideration, even though the industrial systems use the same electronic components as the business systems and are exposed to harsher environments.

RECALLING THE FACTS

1. _____ is the general sequence and arrangement of coded information on a punched tape.
2. The most common type of tape format in use is the _____.
3. The _____ format is the same as the _____ format but without programmed decimal points.
4. The _____, developed years ago by the EIA, combines the best features of both the binary and decimal systems.
5. _____ is the international coding standard for all CNC, information processing, and communication systems.
6. Modern CNCs can automatically detect and adjust for either _____ or _____ input.
7. The primary causes of CNC downtime are _____, _____ disturbances, contaminated _____, _____, and _____.

SECTION 2
REVIEW

KEY POINTS

- N/C data enters the MCU through one of five means: (1) tape readers, (2) cassette tapes, (3) floppy disks, (4) direct line to the CNC, and (5) manual data input (MDI).

- Machine actuation registers are like storage mailboxes within the control that accept the N/C data and pass the input to the motors that move and position the machine slides. Buffer storage is temporary storage that holds data prior to entry to the active machine registers. Buffer storage reduces dwell time between machine moves.

I think I understood all that, but let's go over the highlights.

- CNC machine motion is supplied by drive motors that move and position the moving components of the machine tool. Machine tool drive motors are of four types: stepper motors, direct current (DC) servos, alternating current (AC) servos, and hydraulic servos.

- Rotary motion from the drive motors is converted to linear (straight line) motion by recirculating ball-lead screws called *ballscrews*.

- Electrically driven machines can have table travel speeds in excess of 400 in./min. and as high as 1,200 in./min. or higher for certain types of CNC machines.
- *Resolution* of a CNC machine is the smallest movement the machine is capable of making.
- Feedback systems make sure the CNC machine is positioned exactly where the CNC control thinks it is supposed to be. Feedback systems may be open or closed loop. Virtually all CNC machines today are closed loop.
- CNC machines position themselves along coordinate axes. *Axis* refers to any direction of motion that is totally controlled by specific N/C commands.
- Standing in front of a vertical CNC machine, the X-axis machine movement would be the machine table moving left and right; Y-axis movement would be the table moving in and out (toward you and away from you); and Z-axis movement would be the spindle moving up and down.
- A line through the center of the machine spindle is the Z-axis.
- The entire concept of numerical control is based on the principle of *rectangular coordinates*.
- Rotational movements consist of A-, B-, and C-axes. The A-axis rotates around the X-axis, the B-axis rotates around the Y-axis, and the C-axis rotates around the Z-axis.
- With incremental dimensioning, each dimension is given as an incremental distance from the last position to the next position. Incremental programming works the same way; it positions the cutter from the point where it currently is to the next point programmed.
- With absolute dimensioning, each dimension is given from the same zero location or baseline. Absolute programming works the same way; all positions are figured from the same zero or reference point and all positional moves come from the same "datum edge."
- Tape format is the general sequence and arrangement of coded information on a punched tape.
- The most common type of tape format in use today is decimal point programming.

- The Electronics Industries Association (EIA) developed the binary-coded decimal system (BCD) for coding tapes years ago. The American Standard Code for Information Interchange (ASCII) is another coding system which is now the standard coding system used throughout the computer and communication industries.
- The primary causes of N/C and CNC and other computer equipment downtime are heat, power disturbances, contaminated air, vibration, and oxidation.

REVIEW QUESTIONS AND EXERCISES

1. N/C data enters the MCU through one of five means. These are: floppy disks, tape readers, direct line to CNC, _____, and _____.
2. Explain the function of machine actuation registers.
3. The most common type of tape reader used today is the _____ type.
4. Write a short paragragh describing the value of buffer storage.
5. The rotary motion from the drive motors is converted to linear (straight line) motion by recirculating ball-lead screws called _____.
6. The CNC knows that the machine is properly positioned if the control and machine form a _____ system.
7. In your own words, describe X-, Y-, and Z-axis movements on a vertical machine.
8. A line through the center of the machine spindle is always the Z-axis. (True/False).
9. Name which quadrant the following points are in:
   a. (6, 10)
   b. (2, −5)
   c. (.5, −25)
   d. (−4, 6)
   e. (−10, −16)

10. The B-axis rotates around the X-axis. (True/False)
11. _____ programming positions the cutter from the point where it currently is to the next point programmed.
12. _____ programming always positions the cutter from the same zero or reference point.
13. If a positioning error occurs in an _____ system, all remaining moves will be wrong.
14. "Zero shifting" the machine enables the zero position to be set anywhere on the machine table as long as the machining positions stay within the X, Y, and Z range of the machine tool. (True/False).
15. _____ N/C systems download the entire program into the MCU's memory.
16. _____ is a means of programming the machine tool through the CNC keyboard and push-buttons.
17. The most common type of tape format in current use today is _____.
18. The word address format, standardized by the EIA years ago, uses a _____ to identify each separate word.
19. Name the two purposes that the letter address has for N/C words.
20. The _____ system is now the standard coding used throughout the computer and communication industries.
21. List the primary causes of N/C, CNC, and other computer equipment downtime.
22. List and explain five new words you learned in this section.
23. Using full sentences and in your own words, write a paragraph describing what the MCU is, what it does, and how it basically works.

SHOP ACTIVITIES

1. Draw up a chart to list and describe the types of N/C and CNC machines in your company's or school's shop. Items to include on the chart are: machine type, whether the machine

is vertical or horizontal, control type, how most programming is done (absolute or incremental mode), axis directions for each machine, method of N/C data input (tape, DNC, floppy disk, etc), and how programming is handled (off-line or shop floor, computer based or manual, etc.). Add any other factors that would best help describe the N/C environment for your shop.
2. Identify drive motors on various machines for the different axes.
3. Talk to maintenance. Find out the most common causes of N/C and CNC downtime for your shop. Perhaps maintenance has records which could be shared with classmates describing problems, causes, solutions, and length of time to fix the problem.
4. Review some shop program manuscripts on various machines to determine:
    a. zero position and how machine alignment was obtained
    b. whether absolute or incremental programming was used
    c. operations performed (milling, drilling, tapping, etc.)
    d. how N/C data is input to the MCU
    e. how errors are corrected

SECTION 3
BASIC CNC FUNCTIONS AND FEATURES

OBJECTIVES

After studying this section, you will be able to:

- Identify and explain the various words in a typical CNC block.
- Describe the function and purpose of each CNC word in a block of information.
- List common CNC preparatory functions and explain the operations performed.

X....Y....G....M....F....S....?

INTRODUCTION

In this section you will begin learning the language that will enable you to talk to the CNC and give it action commands to do work. As with learning any language, you must first learn how the language is structured, what the words mean, and how you use them within the language.

The language of CNC is a lot like English. Just as our language is made up of words, sentences, and paragraphs, CNC systems also have a language that uses the equivalent of words, sentences, and paragraphs. For example, you can read a story in English and understand quite clearly what is happening. A CNC program tells a story, too. It tells a story about the machining of a particular part on a CNC machine for a specific operation as the program's codes, functions, and commands guide the cutting tools through the machine moves and motions to cut the part.

As you have already studied in Section 2, a CNC program is made up of blocks which are like sentences because they contain one instruction or one move for the CNC machine. Each block is made up of words, just like English sentences. Punctuation in English is made up of special characters such as periods, colons, semicolons, question marks, and so forth. Punctuation in a CNC program contains some special characters, too. Some of these are similar characters for beginning the N/C data format and others are used at the end of each block of information, as well as for comments within the program.

In this section you will also learn about some of the commands that cause specific operations to be performed. These are called *preparatory functions*. They are also called *G functions* or *G codes*. Some CNC machines use various conversational programming systems, but G codes are still the foundation of CNC programming. Understanding these G codes is very important to your success in this field because they cause machining operations to be performed such as milling, drilling, tapping, etc. The CNC doesn't know or care whether you have selected the proper G code or whether the X and Y positions you programmed are correct; it does what it is told by the program. Remember, as we studied in Section 1, it is up to you to know exactly what to tell the CNC and to check all your work *before* you run the program. After an accident or a wreck happens, it is too late.

UNIT 16

Functions Controlled by N/C and CNC

KEY TERMS

manuscript debug operator information
machine tape information heading

As you have already studied, CNC programming is either done at the machine or off-line, in the office. An operator at the machine will generally begin with paper and pencil to write down some setup ideas and instructions, make a sketch, or do some calculations before beginning to program at the CNC machine. In some cases, he or she may handwrite the entire program out on paper before beginning to enter the CNC data and commands into the MCU.

Off-line programming, regardless of machine type, creates the actual CNC data and the program *manuscript*. The program manuscript is a complete, printed output copy of all the CNC codes, commands, and functions of a CNC program. The program manuscript is a communication tool for the programmer and the operator that helps them "visualize" what is happening at the machine while this program is being run. It also helps to *debug* programming problems or find errors in the part program. Skilled programmers and operators can read and understand a program manuscript and tell what the machine is doing or will do at any particular point in the program. Knowing how to read and understand a manuscript is very important to finding and correcting programming errors or problems. An example of a typical CNC manuscript is shown in Figure 16–1.

A CNC program manuscript consists of the *heading, machine tape information,* and *operator information*. The heading contains specific identification information. Identification information consists of part name and number, drawing and fixture number, and the pro-

O/N SEQ.	G PREP. FUNCT.	X± POSITION	Y± POSITION	Z± POSITION	R± POSITION	I/J/K POSITION	A/B/C P/Q	POSITION ± WORD	F/E FEED RATE	S SPINDLE SPEED	D/M WORD	T TOOL WORD	M MISC. FUNCT.
O182	G00											T16	M 6
O183	G81	X 54650	Y 67949	Z-12500	R 36195		B 180000		F50	S1100		T17	M 3
N184		X 31467	Y 32447										
N185		X- 9551	Y 23335										
N186		X-50553	Y 45953										
N187		X-54165	Y 87051										
N188		X-35147	Y 122653										
N189		X 9551	Y 131665										
N190		X 45053	Y 109047										
N191	G00			Z 36295									
O192	G00											T17	M 6
O193	G84	X 45053	Y 109047	Z- 9375	R 36195		B180000		F254	S 440		T19	M 3
N194		X 54650	Y 67949										
N195		X 31467	Y 32447										
N196		X- 9551	Y 23335										
N197		X-50553	Y 45953										
N198		X-54165	Y 87051										
N199		X-35147	Y 122653										
N200		X 9551	Y 131665										
N201	G00			Z 36295									
O002	G00											T19	M 6

FIGURE 16–1
An example of a typical CNC manuscript.

grammer's name. Machine tape information contains all the codes, commands, and functions necessary for proper machine operation, such as preparatory functions, miscellaneous codes, X, Y, and Z positions, and feed rate and spindle speeds. This information will vary, depending on the particular CNC machine tool being used and the particular part to be machined.

The operator information section of a CNC program contains information that you will be studying more of in this section. This information consists of position number, depth of cut, operation description, and cutting tool information. Also included in this section is information on setup and machine alignment, as well as any other information the programmer feels the operator should know. This information is used strictly as an aid to the operator for a better understanding of how to prepare for and process the job.

The specific words controlling CNC are part of the machine tape information and are printed on the manuscript for operator and programmer reference. These words consist of the following: sequence numbers; preparatory and miscellaneous functions (typically G and M); X, Y, and Z, etc.; coordinate information; spindle speeds; and feed rate. As you have studied, CNC words consist of alphanumeric (combination of alphabet letters and numbers) codes, which relate to a specific register in the MCU and are used to either provide information or cause an appropriate machine tool movement or action to occur. Each of the basic CNC words will be discussed in more detail in the following units.

RECALLING THE FACTS

1. CNC programming is either done at the _____ or _____, in the office.
2. The program _____ is a printed output copy of all the CNC codes, commands, and functions of a CNC program.
3. Programmers must _____ their part programs by finding errors in their part programs and correcting them.
4. The _____ part of the program _____ contains identification information which contains _____, drawing and fixture number, and programmer's name.

5. _____ information contains all the codes, commands, and functions necessary for proper machine operation.

6. The operator information section of a CNC program contains _____ _____ _____.

7. The specific words controlling the CNC are part of the _____ information section of the manuscript.

UNIT 17

Sequence Numbers

KEY TERMS

sequence numbers   *CRT*   *tape search*

Sequence numbers are words that begin with the letter address "N" followed by numbers. Sequence numbers are normally the first word in every block of N/C data and their use is informational rather than functional. The primary purpose of sequence numbers is to:

- identify and sequence each block of information so that it can be distinguished from the rest, and
- indicate the block of information being acted upon by the MCU and which ones are in buffer storage to be acted on next.

Sequence numbers can begin with any number and generally increase by 1s, 5s, or 10s. They are written, "N001, N002, N003" for every block throughout the entire program. The sequence number word, as well as other words, appears on the operator's program manuscript (Figure 16–1) and CNC *CRT* (cathode ray tube) screen (Figure 17–1).

The sequence number is usually a three- or four-digit word. Older controls consisted of only three digits, and sequence numbers could only go as high as 999 without having to start the numbers over again. With newer controls, adding the extra digit now allows sequence numbers to go as high as 9999.

> **"Tech Talk"**
>
> Quality is and should be defined by the customer and is based on buyer preferences, continuous cost improvements, and value.

CNCs are equipped with a *tape search* feature. This feature allows the CNC operator to search the program for a particular sequence number and stop when the number is found. Essentially, the

FIGURE 17-1
The sequence number word (and other words) appears on the operator's CNC CRT screen and allows the operator to identify each block of information and track progress through the program. (Courtesy of GE Fanuc Automation.)

tape search feature allows CNC program information to be temporarily bypassed or recalled as needed.

Modern CNCs can search for any specific block in the part program. Older controls used either an H- or an O-word address along with N. H and O letters for sequence numbers were used for "full" or complete blocks of information; N words were used for all other blocks. H and O blocks were ordinarily used at the beginning of the tape and after every tool change. The reason for this was that many older controls were equipped with tape search capabilities that would only search for H or O blocks. So, if you wanted to search forward or backward in the tape, you could only search for H or O blocks and you could only restart before or after every tool change. H or O blocks also had to be "full information" blocks in order to make sure all machine actuation registers were reloaded with the proper data before a restart. Virtually all CNCs today, regardless of make, use the letter "N" as the word address that precedes the actual number.

RECALLING THE FACTS

1. _____ numbers are words that begin with the letter address "N" followed by numbers. Their primary purpose is to:
   - _____ and _____ each block of information so that it can be distinguished from the rest, and
   - _____ the block of information being acted on by the _____ and which ones are in _____ storage to be acted on next.
2. The sequence number word, as well as other words, appears on the operator's program manuscript as well as the CNC _____ which stands for _____ _____ _____.
3. CNCs are equipped with a _____ feature which allows the operator to _____ the part program for a particular _____ number and stop when that number is found.
4. Virtually all CNCs today, regardless of make, use the letter _____ as the word address that precedes the actual number.

UNIT 18

X and Y Words

KEY TERMS
X and Y words   centerline   cutter radius
zero suppression

PROGRAMMING WITH X AND Y WORDS

Regardless of the type of CNC machine programmed, coordinate information is necessary for the machine tool to position itself. The coordinate information may be expressed using X, Y, Z, and other words. However, in this preliminary discussion you will be studying about *X and Y words*.

Because decimal point programming is the most common format in use today, X-, Y-, and Z-word input normally uses only as many significant digits as required. The X- and Y-word address still needs to be programmed, but insignificant digits may be dropped. For example, if an X dimension of 5.5 inches is required to be programmed, it would be programmed as X5.5. It would not be necessary to program X5.500 or X005.500. These extra zeros are insignificant as far as the MCU is concerned because they serve no purpose and add no greater accuracy to the programmed dimension.

Many CNC machines still use the interchangeable or compatible format. X and Y words in this format are seven-digit numbers, along with the sign of the number, preceded by the letter "X," to indicate the X-axis and the letter "Y" to indicate the Y-axis. The X and Y words are written as X±xxxxxxx and Y±xxxxxxx, where x equals some particular numerical value. The interchangeable or compatible format is not programmed with decimal points; the position of the decimal point is fixed in this tape format to allow four places to the right of the decimal point. The command "X+43750" specifies the dimension of 4.3750 in the X-axis direction and is accepted by the control system. Most controls have the capability of retaining the sign of the programmed word. Because of this, and depending upon the machine and control system, the sign of the number need

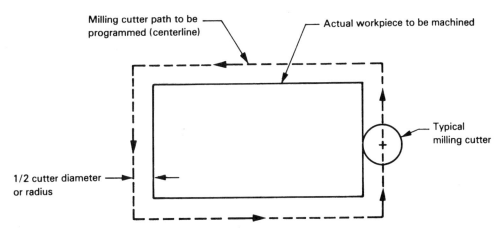

FIGURE 18-1
The cutter radius must always be allowed for when programming X and Y words for milling operations.

only be programmed when it changes from the previous block of information. On older controls with H or O blocks, the sign of the number should always be programmed to ensure the correct sign input when starting a series of operations.

It is important to remember when programming X and Y words that the actual *centerline* of the cutter is always programmed, as shown in Figure 18-1. Calculating X and Y values for point-to-point (hole pattern) operations is easier because the X and Y word input is the centered location of the hole. Milling, however, is more involved since the *cutter radius* must always be allowed for in programming for X and Y (centerline) locations. You need to study and understand the example in Figure 18-1 very well because this principle applies across all CNC machines. Failure to understand or remember this important concept can wreck machines, scrap parts, damage tools, and injure people.

ZERO SUPPRESSION

Older N/C units required a certain number of digits in each block. If a digit was not needed, the space was filled with a place-holding zero. Zeros were placed before and after significant numbers to fill the full block. Today, CNC controllers do not require leading and trailing zeros as decimal point programming is now the accepted standard and decimals are programmed into the words.

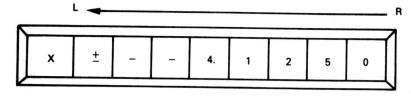

FIGURE 18–2 Coordinate information showing fixed decimal location and right-to-left registration order.

However, some N/C and CNC controllers are still in use using the interchangeable or compatible format and *zero suppression*. The coordinate information in the word-addressed X and Y registers enters, in most cases, in a right-to-left manner, as shown in Figure 18–2. This automatically positions the decimal point four places to the left of the X or Y word.

Study Figure 18–2. While there are seven digit positions available, only five are needed. This means that the word could be written as X+0041250, and the two preceding zeros could be programmed and entered into the control. However, because the words enter the registers from right to left, the first two leading zeros are insignificant. This means they have no bearing on the actual value of the number. In other words, the number is the same whether the first two zeros are there are not. Therefore, they can be suppressed and do not need to be programmed. *Suppress* means to keep from being revealed or to hold back. The word would be acceptable to the control and machine with or without the leading zeros. The leading zeros are insignificant, though, and have no effect on the programmed word. This omission of an insignificant zero digit is called leading *zero suppression*.

RECALLING THE FACTS

1. _____ words are necessary in order for the machine tool to position itself.
2. The interchangeable or compatible format is _____ programmed with decimal points.
3. In decimal point programming, an X position of 7.000 and a Y position of 4.500 could be written as _____ and _____.
4. In the word address interchangeable or compatible format, an X position of 10.5625 and a Y position of 3.0000 would be written as _____ and _____.

5. It is important to remember when programming X and Y words that the actual _____ of the cutter is always programmed.
6. Calculating X and Y positions for hole locations is easy because the actual centered _____ of the hole is used. Milling is more involved since the _____ must be allowed for in programming X and Y _____ locations.
7. Failure to allow for the _____ when calculating X and Y words for milling operations can wreck the machine, damage parts and tools, and injure people.
8. The omission of an insignificant zero digit is called *leading* _____.

UNIT 19

Feed Rates and Spindle Speeds

KEY TERMS

feed rates
F word
constant surface speed

inches per minute
S word

inches per revolution
spindle speeds

FEED RATES

Feed rates control the amount and rate of metal removal for a particular tool and type of workpiece material to be machined. Feed rates are normally measured and programmed in *inches per minute* (ipm) but can also be programmed in *inches per revolution* (ipr) of the spindle, millimeters per minute (mm/min.), or millimeters per revolution of the spindle (mm/rev.).

For most machine tool and control manufacturers, the feed rate word, an *F word*, is programmed directly using decimal point programming. The feed rate code is an F-word address followed by the feed amount. You would program F4.5 for a feed rate of 4.5 inches per minute, for example. Machining centers usually are programmed in inches per minute. Turning centers are normally programmed in inches per revolution. You would program F.007 for a feed rate of .007 inches per revolution. G codes, as you will soon learn, are sometimes used for switching from ipm to ipr or back again within the same program. For example, the G code to switch to ipm feed rate is G94 and the G code to switch to ipr is G95.

Some older CNCs used a format where the decimal point was not programmed but was fixed to allow for one place to the right of the decimal point (Fxxx.x). For example:

- .5 ipm would be programmed as F5
- 30 ipm would be programmed as F300
- 100 ipm would be programmed as F1000

FEED RATES AND SPINDLE SPEEDS • 169

Feed rate selection and use depends on many things. Some of these include:

- the type of machine tool and cutting tool selected
- rigidity of the machine tool
- rigidity of the setup
- material type
- horsepower of the machine tool

The maximum and minimum feed rates per axis of the machine tool will vary depending on the machine and control manufacturer. These feed rates dictate the permissible feed range.

SPINDLE SPEEDS

Spindle speeds, measured in revolutions per minute (rpm), indicate the number of revolutions that the spindle makes in one minute. The spindle speed code is an *S-word* address followed by the rate in rpm. A spindle speed of 350 rpm, for example, would be programmed as S350. The spindle speed is generally governed by the work or cutter diameter, type of cutting tool, and material type. If the spindle speed is programmed in rpm, the spindle will maintain a constant spindle speed until changed.

Spindle speeds in feet per minute (fpm) are also used. Sometimes referred to as *constant surface speed* (CSS), programming in fpm or CSS allows the machine to maintain a constant cutting speed regardless of the part diameter. For example, a facing pass programmed in CSS on a CNC turning center (Figure 19–1) will cause the spindle to rotate faster to keep the cutting speed constant as the cutting tool moves closer to the center of the part. If several rough turns are taken, each pass will be at a slightly higher spindle speed as the diameter gets smaller.

> **"Tech Talk"**
>
> Metals are generally classified into two types; ferrous and nonferrous. Ferrous metals contain iron and will give off a spark when held in contact with a grinding wheel. Ferrous metals are classes of steel and cast iron. Nonferrous metals contain little or no iron. They are resistant to corrosion and are nonmagnetic. The most commonly used nonferrous metals are: aluminum, copper, lead, nickel, tin, and zinc.

FIGURE 19-1
A facing pass programmed in constant surface speed (CSS) on a turning center causes the spindle to rotate faster keeping the cutting speed constant as the cutting tool moves closer to the center of the part. (Courtesy of Cincinnati Milacron, Inc.)

G codes are sometimes used for switching from rpm to CSS and back again within the same program. For example, the G code to switch to the CSS mode is done with a G96 in this block:

N11 G96 S250 M3

Switching to the constant rpm mode is done with a G97 in this block:

N13 G97 S1400 M3

Spindle speed ranges are extremely important to the success of a CNC machine installation. Like feed rate ranges, spindle speed ranges will vary among machine and control manufacturers. But as long as programmers stay within the required spindle speed range, any spindle speed can be programmed. Spindle speeds will be discussed later in the units discussing programming.

RECALLING THE FACTS

1. _____ control the amount and rate of metal removal for a particular tool and type of workpiece material to be machined. They are normally measured and programmed in inches per _____ and inches per _____.
2. When programming feed rates, the _____ word is programmed directly using decimal point programming.
3. In decimal point programming, a feed rate of 7 inches per minute would be programmed as _____. A feed rate of .010 inches per revolution would be programmed as _____.
4. In the word address interchangeable or compatible format, a feed rate of 10 inches per minute would be programmed as _____ and a feed rate of 3.2 inches per minute would be programmed as _____.
5. Feed rate selection and use depends on the type of machine tool and _____ selected, _____ of the machine tool, _____ of the setup, and _____.
6. _____ speeds begin with an _____ word address and are the number of _____ the spindle makes in one minute.
7. An rpm of 750 would be programmed as _____ and an rpm of 1200 would be programmed as _____.
8. Spindle speed on a CNC machine is generally governed by the work or cutter _____, type of cutting tool, and _____.
9. _____ allows the machine to maintain a constant cutting speed regardless of the part diameter.

UNIT 20

Programmable Z Depth

KEY TERMS

Z word   *R word*

Most modern machine tools use a *Z word* to program their Z direction motion. Although programming the Z word may differ among manufacturers, the Z motion must be programmed accurately by the programmer if the machine tool is to produce quality workpieces.

Z motion, as you may recall, on a vertical or horizontal CNC machine for example, is the motion of the Z-axis into or away from the work. A positive Z move (+Z) is a move of the spindle away from the workpiece. A negative Z (−Z) is a move of the spindle into the workpiece. Z-coordinate input, like X and Y words in decimal point programming, uses only as many significant digits as required. The Z-word address still needs to be programmed, but insignificant digits may be dropped. For example, if a Z depth of 2.5 inches is required to be programmed, it would be programmed as Z − 2.5.

Some older CNCs used a seven-digit word preceded by a plus or minus sign and the Z-word *address*. The Z motion is relative to a "rapid-to-position" move of some type, where normally the spindle will rapid to a point close to the part, usually within .100 inch, or so, before beginning to feed the tool into the workpiece. This rapid-to-position move is normally accomplished through an *R word* (to be discussed later) or a rapid traverse fixed rate with a specific Z.

If only the Z-word is programmed to control the entire Z-axis movement, the following formula can be used in most cases:

$$Z = PS + CL + TL$$

where:

Z = distance from Z0 to spindle gauge line
PS = distance from Z0 to the part surface
CL = clearance (if needed)
TL = tool set length from spindle gauge line to cutting edge*

To find the Z value for the part and tool in Figure 20–1, the following calculation is made:

$$Z = PS + CL + TL$$

where:

PS = 4.0000

TL = 6.0000

CL = 0.0000 (cutter positional to part surface)

---

*Since many systems are arranged with a tool length storage feature, the control in some cases will add the tool set length (TL) to the programmed Z value. In this case, the programmer need not include the tool set length dimension when calculating the Z word.

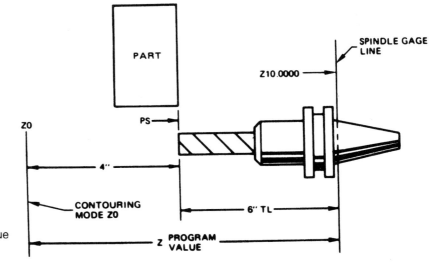

FIGURE 20–1
Part and tool for Z value calculation example.

$$Z = 4 + 0 + 6$$
$$Z = 10.0000$$

You will study and learn more about Z motion later when programming examples are discussed.

RECALLING THE FACTS

1. Most modern CNC machine tools use a _____ to program their Z-direction motion.
2. To move the cutter into the part, a _____ would be programmed; to move the cutter away from the part, a _____ would be programmed.
3. To program a depth of 4.8 inches in decimal point programming, a _____ would be programmed; to program a depth of 10.2 inches in decimal point programming, a _____ would be programmed.
4. On some CNCs, Z motion is accomplished through an _____ where the _____ represents a "rapid to" point close to the part where the cutter will rapid to before it begins to feed.
5. More practice with coordinates. Fill in X, Y, and Z values.
    a.

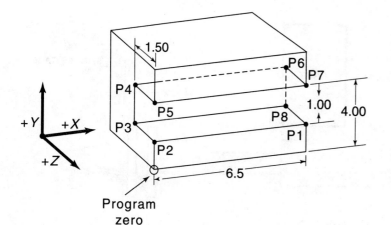

Point	X	Y	Z
P1	6.50	1.00	0
P2			
P3			
P4			
P5			
P6			
P7			
P8			

b.

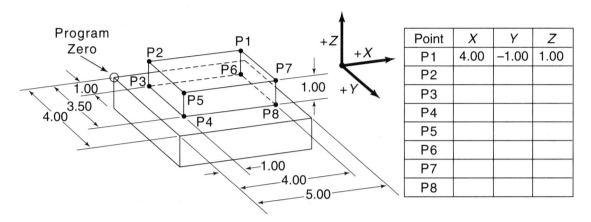

6. Programming Z depths. Fill in blocks with 4 Z programming depths. In which quadrant is this part programmed?

   a.

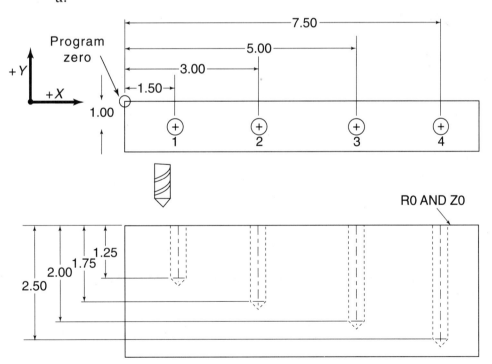

- Operation is to drill a .375 hole to the depths shown.
- **Remember!** Add the measurement of the drill point to get the required print depth. To do this, multiply .3 × the drill diameter and add this value to print depth (.3 × .375 = .113). Therefore, a 1.00″ deep hole with a 3/8″ diameter drill should be programmed to Z − 1.113.

N10	T1	M06			Tool change to tool 1
N20	S800	M03			800 RPM, spindle on CLW
N30	G0	X1.5	Y − 1.0		Rapid traverse to hole 1
N40	G81	R.1	Z	F3.	Rapid Z to .100, drill to Z depth at 3 IPM
N50	X.3	Z			Canned cycle drill at hole 2
N60	X.5	Z			Canned cycle drill at hole 3
N70	X7.5	Z			Canned cycle drill at hole 4
N80	G80	M9			Cancel G81 drill: Turn off coolant
N90	G0	G91	G28	Z0	Tool rapid traverse to Z home position
N100	M30				Machine stops, M codes cancelled, and program resets back to beginning

b.

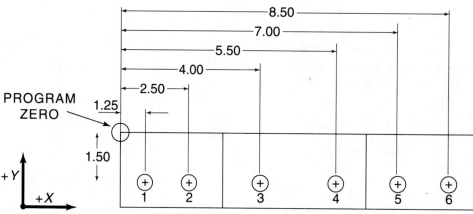

- Operation is to drill a .375 hole to the depths shown.
- Remember to calculate for the drill point.
- Add .020 − .050 inches to the depth of a through hole to allow the burr that forms when the drill exits the part to be cut off.

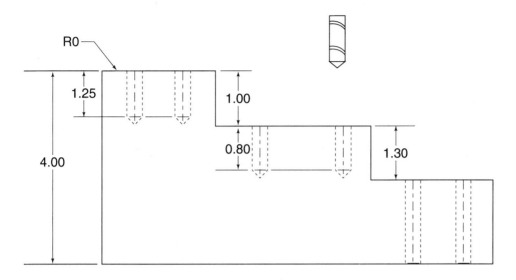

More programming Z depths. Fill in blocks with 3 Z programming depths. In which quadrant is this part programmed?

- Using a .375 drill, bore 3 places to the depths shown.
- Remember to calibrate allowing for the measurement of the drill point.
- Add .20 – .50 inches to the depth of a through hole to allow the burr that forms when the drill exits the part to be cut off.

N10	T1	M06				Tool change to tool 1
N20	S800	M03				800 RPM, spindle on CLW
N30	G0	X1.25	Y – 1.50			Rapid traverse to hole 1
N40	G81	R.1	Z	F3.0		Rapid Z to .100, drill to Z depth at 3IPm
N50	X2.5					Canned cycle drill at hole 2
N60	X4.	R – 1	Z			Rapid Z – 1.00 from R0 – drill hole 3
N70	X5.5					Canned cycle drill at hole 4
N80	X7.0	R – 2.3	Z			Rapid Z – 2.3 from R0 – drill hole 5
N90	X8.5					Canned cycle drill at hole 6
N100	G80	M9				Cancel G81 drill: Turn off coolant
N110	G0	G91	G28	Z0		Tool rapid to Z home position
N120	M30					Machine stops – program resets

UNIT 21

Miscellaneous Functions

KEY TERMS

miscellaneous functions M06 M00
M01 M word modal
M30 nonmodal

The word *miscellaneous* means "made up of a variety of parts or ingredients; having a variety of characteristics, abilities, or appearances." *Miscellaneous functions* in CNC, just like the definition states, perform a variety of functions and commands. Generally, they are multiple-character, on/off codes that determine a function controlling the machine. Miscellaneous functions are special features called to use by an *M word*. M words are functional at the beginning or end of a cycle and are two-digit numbers preceded by the letter "M" (Mxx). They activate "special" functions such as spindle start, coolant control, program stop, and others.

The following is a list and explanation of some basic miscellaneous functions in accordance with EIA standard coding:

- *M00 (Program Stop)*. This code stops the reading cycle after the movement or function has been completed in the block in which the program stop was coded. In addition, this code will also turn off the spindle and coolant if it is turned on.

- *M01 (Optional Stop)*. This code, like the (M00) program stop, stops the reading cycle after the movement or function has been completed in the block in which the optional stop was coded. This code will also turn off the spindle and coolant if it is turned on. However, the code will only function if the operator has selected optional stop on the control. If no optional stop has been selected, the M01 will be read but no action stop will occur.

- *M30 (End of Program)*. After the movement or function has been completed in the block in which the (M30) end of program was coded, this code will stop all X, Y, and Z motion and turn off the spindle and coolant. In addition to stopping the spindle and coolant and stopping any further motion, the (M30) end of program will reset the programmed input to begin again or rewind the tape to the leader (front) portion of the tape.
- *M06 (Tool Change)*. This function should be coded in the last block of information in which a given tool is used. The specific machine tool design determines the sequence of events during a tool change (Figure 21–1). This code also stops the spindle and coolant, if turned on, and retracts the tool to the full retract position for the tool change.

FIGURE 21–1
A tool change as generated by the M06 tool change miscellaneous function. (Courtesy of Cincinnati Milacron, Inc.)

There are other miscellaneous functions besides these, but these are some of the primary ones for you to be familiar with.

Commands may be either single, one-at-a-time commands or commands that stay in effect until changed by another command or canceled. Single one-at-a-time commands are called *nonmodal* commands. A tool change (M06) is an example of a nonmodal command used to execute a single tool change at the end of a CNC block or instruction. Commands that stay in effect until changed or canceled are said to be *modal* commands. A feed rate (F5.5) is an example of a modal command. The feed rate remains the same until it is changed by another feed rate later in the program.

RECALLING THE FACTS

1. _____ functions are multiple on/off codes that determine a function controlling the machine.
2. _____ words are functional at the beginning or end of the cycle and are two-digit words preceded by the letter address _____.
3. _____ words activate special functions such as _____ start, coolant control, and _____ stop.
4. The _____ word for program stop is _____, for optional stop is _____, for end of program is _____, and for tool change is _____.
5. A single, one-at-a-time command is called a _____ command. An example of a _____ command is _____.
6. Commands that stay in effect until changed or canceled are called _____ commands. An example of a _____ command is _____.

UNIT 22

Preparatory Functions and Operations Performed

KEY TERMS

preparatory function   G code   linear moves
G00   G01   canned cycle
schematic

PLANNING MACHINE ACTIONS

The word *preparatory* means "to make ready or prepare; introductory." The **preparatory function** in CNC prepares and introduces different actions or operations to occur on the machine.

The preparatory function or cycle code in CNC is a two-digit number preceded by the word address letter "G" (00–99). This code, referred to as the *G code*, determines the mode of operation for the machine. There have been attempts to standardize G codes among CNC manufacturers and many are common and standardized to some degree, but nothing officially has been finalized.

There are four main groups or categories of G codes. They are:

1. to select a movement system (at a programmed feed rate or a rapid move),
2. to select a measurement system (metric or English),
3. to program for compensation and differences in tool lengths and diameters, and
4. to select a preset sequence of events known as *canned cycles*.

Linear moves involve straight-line movements of a machine tool. A straight line produced by the movement of two or more axes at

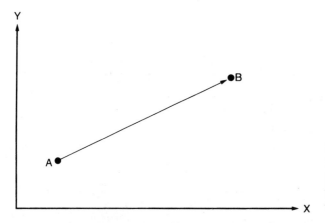

FIGURE 22-1
Linear interpolation between points A and B would cause the X and Y axes to move at the same time.

the same time is called *linear interpolation* (*interpolation* means finding a value between two given values). *G00* and *G01* are the two primary commands used for linear moves on most CNC equipment. G00 is used for point-to-point positioning at the maximum rapid traverse feed rate of the machine tool. G01 is used to position two or more axes at the same time at a programmed feed rate (Figure 22–1).

A *canned cycle* is a combination of machine moves resulting in a particular machining function such as drilling, milling, boring, and tapping. By programming one cycle code number, as many as six, seven, or more machine actions or movements may occur. These seven or more movements would normally take several blocks of programming to accomplish without canned cycles. Most control manufacturers today have both canned and noncanned cycles as part of their standard control package.

PROGRAMMING WITH CANNED CYCLES

Now it is important for you to study some simple and practical programming examples using the words you learned in this section.

Consider a canned cycle that consists of the following sequence of six operations:

1. Positioning along the X- and Y-axes
2. Rapid traverse to point R
3. Drilling

4. Operation at bottom of hole (if any)
5. Retraction to point R
6. Rapid traverse to initial point

Figure 22–2 is a *schematic* (a diagram) of this sequence. An example of N/C code that could be used to program this sequence is as follows.

>N001 G55 G90 X0 Y0
>N002 G43 H2 Z.5
>N003 M03 S1000
>N004 G81 R.05 Z − .1875 F20

In the second line of code, Z.5 sets up the control point at a Z height of 0.5 inches (that is, 0.5 inches is the first Z position called). In the fourth line of code, R.05 establishes a predetermined "rapid to" point just above the work surface. The value of R is changeable on some controls and fixed at 0.100 inches on others. The R word will be discussed in more detail later in the text.

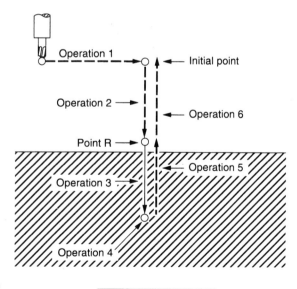

FIGURE 22–2
A canned cycle consisting of a sequence of six operations.

> **"Tech Talk"**
>
> The quality of the end product and how it is viewed by the customer is the sum total and direct result of the internal attention paid to quality.

The codes G98 and G99 are used to specify the retraction point of the cutting tool after the desired Z depth is reached (Figure 22–3). If neither G98 nor G99 is specified on the canned cycle line of information, initial point return is performed. If return to point R is desired, G99 must be programmed on the canned cycle line of information.

Some basic canned cycles with programming examples are shown in Figures 22–4, 22–5, 22–6, and 22–7.

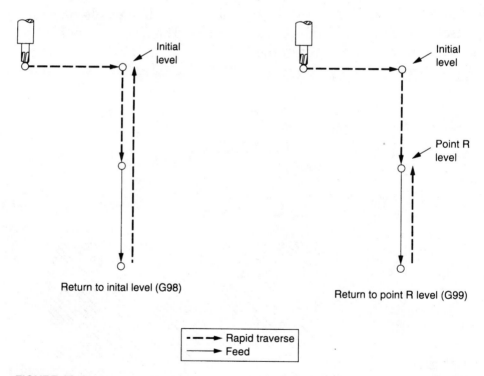

FIGURE 22–3
The codes G98 and G99 specify the retraction point of the cutting tool after the desired Z depth is reached.

G81: Center-drilling, or Spot-drilling, Cycle

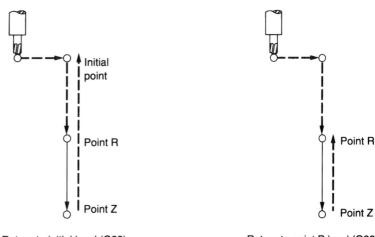

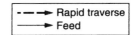

FIGURE 22–4
Center drilling or spot drilling cycle.

G82: Drilling and Counterboring Cycle

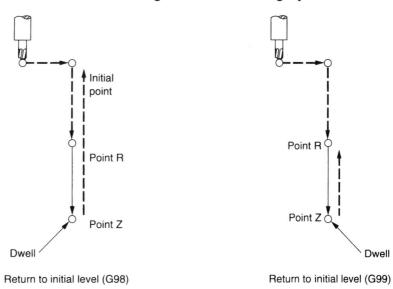

FIGURE 22–5
Drilling and counterboring cycle.

186 • UNIT 22

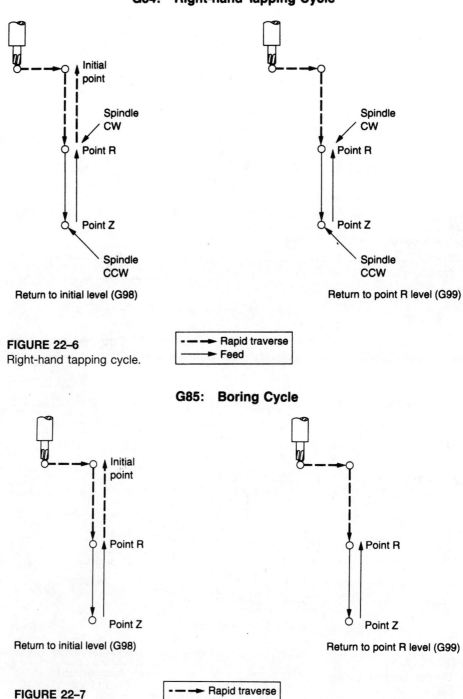

FIGURE 22–6
Right-hand tapping cycle.

FIGURE 22–7
Boring cycle.

RECALLING THE FACTS

1. The _____ function prepares and introduces different actions or operations to occur on the machine.
2. The _____ is a two-digit number preceded by the letter address _____ that determines the mode of operation for the machine.
3. There are four main groups or categories of G codes. They are:
   a. to select a _____ system (at a programmed feed rate or rapid move),
   b. to select a _____ system (metric or English),
   c. to program for compensation and _____ in tool lengths and diameters, and
   d. to select a preset sequence of events known as _____.
4. A _____ is a combination of machine moves resulting in a particular machining function such as drilling, boring, and tapping.
5. Most control manufacturers have both _____ and _____ cycles as part of their standard control package.
6. _____ moves involve straight-line movements of a machine tool; _____ and _____ are the two primary commands used for _____ moves on most CNC equipment.
7. By programming one _____, as many as _____ machine actions or movements will occur. These _____ movements would normally take several _____ to accomplish without _____ cycles.
8. A _____ is another name for a diagram.

SECTION 3

REVIEW

KEY POINTS _____

I'm starting to really understand CNC programming! But let's review anyway.

- The program manuscript is a complete printed output copy of all the CNC codes, commands, and functions of a CNC program; knowing how to read and understand a manuscript is very important to finding and correcting programming errors and problems.
- A CNC program manuscript consists of the heading, machine tape information, and operator information.
- Sequence numbers are three- or four-digit words that begin with the letter address "N" followed by numbers. Sequence numbers are normally the first word in every block of N/C data and their use is informational rather than functional.
- The tape search feature on CNC controls permits searching by sequence number for any specific block in the part program.
- Modern CNC X-, Y-, and Z-word input is programmed using the decimal point and normally uses only as many significant digits as required.

- When programming X and Y words, the actual centerline of the cutter is always programmed. This is particularly important when programming milling operations because an allowance for the cutter radius must always be made.
- The feed rate (F) word is programmed directly using decimal point programming in inches per minute (ipm), inches per revolution of spindle (ipr), millimeters per minute (mm/min.), or millimeters per revolution of the spindle (mm/rev.).
- Spindle speeds, measured in revolutions per minute (rpm), indicate the number of revolutions that the spindle makes in one minute. Spindle speed is an S-word address followed by the rate in rpm.
- Z-coordinate input in decimal point programming uses only as many significant digits as required. A positive Z move (+Z) is a move of the spindle away from the workpiece. A negative Z move (−Z) is a move of the spindle into the workpiece.
- Miscellaneous functions (M functions) are on/off functions controlling the machine. They activate a variety of special functions such as spindle start, coolant control, and program stop.
- Single one-at-a-time commands are called *nonmodal* commands; commands that stay in effect until changed or canceled are called *modal* commands.
- Preparatory functions, or G codes, determine the mode of operation of the CNC such as drilling, tapping, boring, etc.
- A canned cycle is one G code, such as for drilling, for example, that causes a combination of machine moves to occur. Programming one canned cycle (G code) would normally take several programming blocks to accomplish without canned cycles.

REVIEW QUESTIONS AND EXERCISES

1. Describe how the program manuscript helps the CNC programmer and operator.

2. A CNC program manuscript consists of the _____, machine tape information, and operator information.
3. Describe how sequence numbers are used to help the CNC programmer and operator.
4. The _____ feature of a CNC allows the operator to search the program for a particular sequence number and stop when the number is found.
5. When programming X and Y words for milling operations, you must allow for the cutter _____.
6. Explain what *zero suppression* means.
7. Feed rates are normally programmed in _____ or _____ per minute or per revolution of the spindle.
8. A decimal point programmed feed rate of F10.5 ipm is _____ inches per minute.
9. Feed rate selection and use depend on:
   a. the type of machine tool and cutting tool selected.
   b. the rigidity of the machine tool.
   c. the rigidity of the setup.
   d. the type of material.
   e. a and c
   f. all of the above
10. A programmed spindle speed of S1500 is _____ revolutions per minute.
11. To drill a hole to a depth of 3.25 inches, a Z word of _____ would need to be programmed.
12. _____ functions activate "special functions" such as spindle start, coolant control, program stop, and others.
13. Single one-at-a-time commands are called _____ commands. Commands that stay in effect until changed or canceled are called _____ commands.
14. There are four main groups or categories of G codes. They are: (1) to select a movement system, (2) to select a mea-

surement system, (3) to program for compensation and differences in tool lengths and diameters, and (4) _____.

15. The two primary commands used for linear moves on most CNC equipment are _____ and _____.

16. List and explain five new words you learned in this section.

17. Using full sentences and in your own words, write a short paragraph about canned cycles; what they do, how they work, and their value and importance.

SHOP ACTIVITIES

1. Review different shop manuscripts. Identify heading, machine tape, and operator information. Pay particular attention to the specific words controlling the CNC relative to the manuscript you are looking at. What do you notice about the manuscripts that are the same? What differences do you notice?

2. As you review some CNC manuscripts in the shop, how is setup information described to the operator? Are setup drawings included with the manuscript or are simple word descriptions used to tell the operator how to set up the machine? How is the machine aligned and how are setup positions communicated to the operator?

3. With a program entered in a CNC controller, what words are displayed on the CRT screen? What N/C data is displayed relative to the command position (where the CNC is presently)? What N/C data is shown relative to the next position the machine will be moving to? Watch the CRT screen and follow along with the manuscript as you or the operator progress through the program.

SECTION 4

PUTTING IT ALL TOGETHER—SIMPLE PART PROGRAMMING METHODS AND EXAMPLES

OBJECTIVES

After studying this section, you will be able to:

- List and describe the important steps involved in setting up and running a CNC job.
- Identify and program basic CNC lathe and machining center words.

Where and how do I start?

INTRODUCTION

This is a very important section—for several reasons. First, it brings together all of the CNC material you have learned so far. All of the chapters before this one have been building your knowledge base about CNC in order for you to understand what it is, what it does, and how it works, along with CNC's importance and value to manufacturing and to you. This section will enable you to see how all of the material you have learned so far is used.

Second, now that you have this CNC knowledge base, you can begin to learn simple CNC programming; specifically, for a CNC lathe and machining center, since these are the two most common types of CNC machines in use today.

Third, you'll learn the fundamentals of planning and running the CNC job, also very important to successful CNC programming and operation.

UNIT 23

Setting Up the Job and Getting Ready to Run

KEY TERMS

engineering design	manufacturing engineering	single block mode
inventory control	process plan	tool design
setup instructions	revision	check-out
prove-out	dry run	plot

DESIGNING AND PLANNING THE JOB

Before discussing the details of setting up the job and getting ready to run, you first need to understand the high-level, general flow of the part manufacturing process. As you study the flow diagram shown in Figure 23–1, you will see that the process begins in engineering design. *Engineering design* creates product designs and detail drawings for manufacturing and assembly.

The output from engineering design is the part prints to manufacture the components for the company's products. Part prints are typically sent to *manufacturing engineering* where manufacturing engineers plan how the parts are to be manufactured. At the same time, part prints are sent to *inventory control* where the number of parts to be made is determined, how many lots or batches of parts are to be dispatched to the shop at one time, what the schedule or priority will be for those parts to be manufactured, and so forth.

Manufacturing engineering creates a *process plan* for each part that will be manufactured in the plant. Some parts, however, are commercial, which means they are bought from other companies and are not manufactured in-house. The process plan (Figure 23–2) is a description of what needs to be done to complete the manufacturing of the part. It is a document, sometimes consisting of both text and graphics, that describes the methods, machines, sequences, fixturing, and tools to manufacture a detail part. The process plan

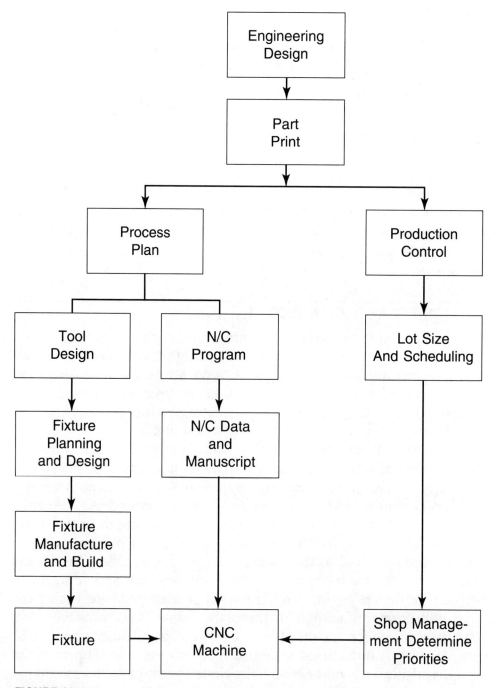

FIGURE 23–1
The part manufacturing process.

Part No	Oper No	Material	SHEET
486973	10	1018	1 of 3
Part Name		**Workstation**	**Planning Location**
Bracket		9672	C3
Operation Name		**Oper. Rev.**	**CVP Operation**
Mill, Drill, Ream, Bore, Tap, Complete		A-3	Yes
Approved by Product/Process Owner			**Production Status**
R.V. Landhoffer (signature)			

Oper. Rev.	Description	Comments
001	Load fixture AB-486973 on machine table. Clean fixture bottom and machine table well before loading fixture.	Check fixture orientation to machine table. 0 on fixture should align with 0° on machine table.
002	Deburr parts and clean fixture before loading. Break all sharp edges .005"/.015". Load and secure part in fixture.	Do not over-tighten clamps!
003	To establish the machine home position on the initial setup, start the first tool to be touched off from the machine home position of Z=0. After machine positions for its first touch-off, adjust the home position to allow for the smallest deviation (in .100" amounts).	Communicate this information to all shifts!
004	Mill, drill, ream, bore, tap complete per CNC program. Program should leave .100" finish stock on 2.0000, 3.5000, and 4.2500 diameter holes. .0625" finish stock should be left on surfaces A, D, and F.	Use proper tooling called for in CNC program. Do not substitute tools!

FIGURE 23-2
The process plan is a description of what needs to be done to complete the manufacturing of the part.

needs to be created in order for manufacturing to know what operations are to be done on what machines and in what order the operations are to be completed.

After the process plan is completed, one copy of the plan generally goes to *tool design*, if a fixture is needed, and another copy goes to N/C programming in order to get the programs started for those operations requiring N/C machining. Tool design is responsible for designing the fixture and tracking the fixture manufacture, assembly, and inspection. Following the tool design flow line in Figure 23–1, you can see that after the tool design department designs the fixture, it must be manufactured, built, inspected, and then stored until needed in production. Following the N/C programming flow line, you can see that the N/C programming department creates the N/C program and manuscript and then either stores the program and manuscript until needed or sends it directly to the shop for the appropriate CNC machine. Shop management decides the priority of the parts to be run and, ultimately, the fixtures, parts, tools, and programs all come together at the CNC machine tool when the part is ready to be run.

> **"Tech Talk"**
>
> Just-in-time (JIT) is a stockless production manufacturing approach whereby only the right parts, both purchased and manufactured, are usable, completed, and available at the right time and in the right place. JIT is about organizing the production process so that inventory of raw materials, work-in-process, and finished goods are driven to a minimum and kept under control.

ELEMENTS OF THE MANUFACTURING RUN

Now let's look at the major factors involved in setting up and getting ready to run the job. When planning to run any CNC machine, you need to consider:

- *Setup instructions.* Along with preparing the N/C program and manuscript to be sent to the shop, the programmer must determine how the part is to be set up on the machine tool, what the alignment or program zero or reference positions/points are, and how to hold and orient the part at the machine. The off-line programmer communicates to the operator how this is to be done by means of *setup instructions* which he or she also prepares in addition to the program itself. Setup instructions may be written (text), pictures (graphics), or both. Making sure setup instructions are clear and understandable to the operator is important for two reasons:

1. to make sure the program matches the part as it was programmed as well as to obtain proper quality and accuracy in the part, and
2. to save time in setting up and completing the job and avoid accidents.

- *Part prints and process plans.* Programmers and operators, most importantly, must check part prints and process plans against the N/C program to make sure that all agree and are the latest *revision.* As parts are changed by engineering, new prints are released and manufacturing changes must be made to process plans, N/C programs, and so forth to suit the changes made to the changed part design. Consequently, revision numbers or change notice numbers are issued and attached to specific part numbers. It is not uncommon for a part to be changed many, many times during the course of its active life in a product. This is a very important concept, and failure to make this all-important check could result in an out-of-date program being run, thereby causing scrap, possible injury, or a wrecked machine or broken cutting tool. Things are constantly changing. Part prints are being modified, process plans are being changed, and programs are getting changed. Make sure the part print is always the latest revision by way of the engineering change notation on the print, the process plan has been updated to the current change, and the N/C program reflects these changes before beginning the job. If in doubt, always ask your supervisor. Part prints, process plans, and setup instructions should always be thoroughly reviewed before running the job in order to familiarize yourself with what the CNC machine tool will be doing with respect to a particular operation and part program.
- *Holding fixtures.* Many N/C programs require a holding fixture to accurately locate and hold the part, as seen in Figure 23-3. As you have already learned, holding fixtures and the machine table should be thoroughly cleaned (free of cutting chips, oil, dirt, and grit) before mounting. This includes the

FIGURE 23–3
Typical holding fixture to hold the part for N/C machining. (Courtesy of Cincinnati Milacron, Inc.)

bottom of the fixture and locating keys. The setup instructions will normally tell you how to orient the fixture so it does not get loaded and clamped 90 or 180 degrees out of position on the machine table, for example. Fixtures to hold raw castings have adjustable locators. If the part has some machining done already, a secondary holding fixture is used to locate the part. This type of fixture will sometimes have precision dowel pins, pads, or other locating devices in order to hold each part in exactly the same position for machining. After the part is located in the fixture or vise, you must determine the starting or zero point for the program. Some fixtures may have a pin or bore to align the machine and program with the part. The method used and the accuracy required will depend on the part's tolerances.

- *Cutting tools*. All cutting tools must be thoroughly checked and examined by the operator in order to make sure they are sharp and tightly held in their holders before beginning to run the part. This can be done as the individual tools are loaded into the tool drum or prior to loading (Figure 23–4). In many shops, cutting tools are assembled by tool setup personnel, and as human error does exist, it's important that another person examine each tool prior to use to make sure it is sharp and tightly held in its holder and driver. All carbide inserts should be checked to make sure the insert has been indexed to a fresh cutting edge and the insert is clamped tightly in its holder.

- *Setting offsets*. Once the starting point of the program is located, the tooling offsets for the machine must be set. The offsets allow the program to move to the fixture, instead of having to move the fixture to the spot where the program is.

- *Check-out and prove-out*. **Check-out** consists of checking the output manuscript, line by line, for any glaring mistakes, problems, or omissions. Check-out is usually done in an office and is sometimes a time-consuming task and often overlooked because programmers feel that all the mistakes can be found in the prove-out stage—on the machine when mistakes are the easiest (and costliest) to find. However, the check-out phase can reveal location problems, feed rate problems, wrong G codes, and so forth, thereby saving time at the machine in the

FIGURE 23–4
Individual tools being loaded into the machine's tool matrix. (Courtesy of Cincinnati Milacron, Inc.)

prove-out phase. ***Prove-out*** generally consists of the programmer being out in the shop at the machine with the operator checking out a new program the first time it is run on the machine. The amount of time spent on a prove-out will depend primarily on four things:

1. the length of the program,
2. the care taken during the preparation of the program,
3. the care taken during the setup, and
4. the care taken during check-out.

Prove-Out: Inspecting the First Part

The purpose of prove-out is to find and correct all major and minor part programming problems on the first part run so that remaining parts will be correct and run within tolerance (Figure 23–5). After the first part is run, it will have to be inspected very carefully. All locations, sizes, depths, diameters, surface finishes, etc. have to be checked to print specifications for accuracy. As a result of the first part inspection, adjustments or changes may need to be made to the part program, cutting tool lengths, tool offsets, fixture offsets, bore size adjustments, or other changes or modifications. A prove-out should always be done to any new program before running continuous parts. Prove-out can be accomplished in several different ways,

FIGURE 23–5
It is important to prove-out any new part program in order to determine any programming errors or part inaccuracies before the program is run in production. (Courtesy of Mazak Corp.)

some of which are more effective than others and may or may not be included as part of the on-machine prove-out. These include:

1. *Making a printed plot of the cutter path.* A ***plot*** is a printed outline of the cutter path on paper indicating rapid and feed cutter path moves relative to the part surfaces. Plotting is generated from a computer output file and is good to obtain general cutter path data but will not show tool depth, feed rate problems, interference problems, etc.

2. *Running a graphics plot of the tool path on the CRT relative to the part surfaces.* This assumes that the programmer has programmed the part using a graphics programming package with tool path verification. This visual verification is also only good to obtain general cutter path data and will not show tool depth, feed rate problems, and interference problems similar to a printed plot.

3. *Making a dry run on the machine.* A ***dry run*** means that the *program* is loaded into the control and executed but no *part* has been loaded. Cutting tools and the holding fixture, if one is used, are loaded and the programmer who programmed the job is typically present with the operator when the prove-out is being conducted. The feed rate override switch on the control may be turned down to slow all the feed rates down in order for the programmer and operator to more easily watch the moves and positions and detect problems or errors.

4. *Running the program in single block mode.* A switch can be set on the control which only lets the program execute one block at a time; that is, run in ***single block mode.*** The cycle start button must be pushed after each block to execute the next block of N/C data. All tools and the part are loaded at this point. The advantage of this prove-out is that all tools and the part are loaded and the program is executing only one block at a time. The feed rate override switch on the control can be turned down to enable the programmer and operator to watch and check all moves and positions and thereby detect problems and errors. This a very effective prove-out method and will determine most of the programming, tooling, interference, and clearance problems.

5. *Running the same part in continuous cycle and at 100 percent feed rate.* This is usually done with the same first part even though the tools are, for the most part, "cutting air" (as this part has already been cut in single block mode). This part of the prove-out is typically done after any programming changes have been made to the program and the part is very close to final acceptance by shop management.

In some cases, several of these prove-out methods are done in series before a particular part is run. For example, a plot may be run first to check all tool paths, followed by a "dry run." Then, the part may be loaded and the single block cycle executed with reduced feed rates. Following this, the same first part may be run after program corrections have been made in continuous cycle and at 100 percent of the programmed feed rate. Once the program has been accepted by shop management and the part has passed inspection, remaining parts in this production run can be completed.

Throughout this entire process, it is very important that you plan a safe prove-out and production run and stay alert and focused on the task at hand. Some general rules to follow are:

- Always know where the master or emergency stop button is on the control as well as the feed hold or cycle interrupt button.
- Stay focused and alert during and after the prove-out. Often a perfect setup will work loose after the forces of machining begin to take their toll on the setup and equipment. Remember, heat, cutting forces, and vibration can cause changes and new conditions to a workpiece setup.
- Pay particular attention to dull or loose cutters, loose chucks, bolts, fixtures, flawed workpiece materials, lost pressure in air or hydraulic chucks, loading parts, and using clamps correctly.
- Do secondary operations, such as burring parts, during the cycle only if there is time.
- Make sure coolant is getting to the cut.
- Safely clear chips from the cut and setup during the run.
- Perform routine maintenance. Make sure the machine is lubricated properly and at the recommended times and coolant is maintained at the proper levels.

- Keep safety guards in place.
- Do not play music in your work area. It is distracting, prevents you from hearing critical machining-related problem sounds, and may be bothersome to your co-workers.
- Do not leave your machine area when the machine is running.
- Check the setup and tools thoroughly at each shift change.

RECALLING THE FACTS

1. _____ creates product designs and detail drawings for manufacturing and assembly.
2. _____ is where manufacturing engineers decide how parts are to be manufactured.
3. _____ decides the number of parts to be made, how many lots or batches of parts are to be dispatched to the shop at one time, and the schedule or priority for the manufacture of those parts.
4. _____ is responsible for designing the fixture and for tracking the fixture manufacture, assembly, and inspection.
5. The off-line programmer communicates to the operator how the part is to be set up on the machine tool by means of _____.
6. Programmers and operators must always check part prints and process plans to make sure they are all at the latest _____.
7. All cutting tools must be thoroughly checked and examined by the operator to make sure they are _____ and tightly held in their holders before beginning machining.
8. _____ consists of checking the output manuscript, line by line, for any mistakes, problems, or omissions.
9. _____ consists of checking out a new program the first time it is run on the machine.
10. The amount of time spent on a prove-out will depend on four things:
    a. _____

b. _____
  c. the care taken during the setup
  d. _____
11. A _____ is a printed outline of the cutter path on paper indicating rapid and feed cutter path moves relative to the part surfaces.
12. A _____ means that the *program* has been loaded into the control and executed but no *part* has been loaded.
13. Through the use of _____, only one block of a CNC program can be executed at a time. The cycle start button must be pushed each time to execute the next block of data.

UNIT 24

Simple CNC Lathe Programming

KEY TERMS
turning centers
inside diameter (ID)
negative X (−X)
outside diameter (OD)
negative Z (−Z)
positive X (+X)
positive Z (+Z)

TYPES OF CNC TURNING CENTERS

In this unit, you will learn some basic CNC lathe programming. Many details relative to tool holders, offsets, compensation, and so forth have been omitted in order for you to understand and grasp the basics of manual CNC lathe programming. Details can be added later after you understand the basics of CNC lathe programming.

Now commonly referred to as *turning centers* because of their increased flexibility and capability, CNC lathes are classified into two types; vertical and horizontal. Vertical CNC turning centers (Figure 24–1) are modern versions of the manual vertical turret lathes (VTLs). Although principally designed for larger and difficult-to-handle workpieces, vertical turning centers have advanced considerably in terms of state-of-the-art CNC features and technology. Other than enhanced features and added technology, the basic construction of vertical turning centers has been, for the most part, fundamentally unchanged.

Horizontal CNC turning centers (Figures 24–2 and 24–3) have changed relative to advanced features and technology, but also in basic construction as well. Many modern horizontal CNC turning centers are of the slant-bed design (Figure 24–4); some with *outside-diameter (OD)* and *inside-diameter (ID)* tools mounted on the same indexable turret. Advantages of the slant-bed design include: easy access for loading, unloading, and measuring; allowance for chips to fall free; minimum floor space utilization; ease and quickness of tool changes; and better strength and rigidity.

SIMPLE CNC LATHE PROGRAMMING • **209**

FIGURE 24–1
A vertical CNC turning center. (Courtesy of Giddings and Lewis, Inc.)

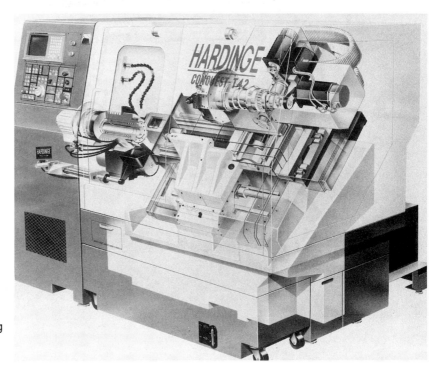

FIGURE 24–2
A horizontal CNC turning center. (Courtesy of Hardinge, Inc.)

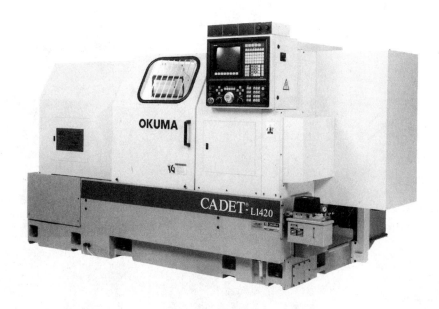

FIGURE 24–3
Many CNC turning centers have changed relative to advanced features and basic construction. (Courtesy of Okuma America Corporation.)

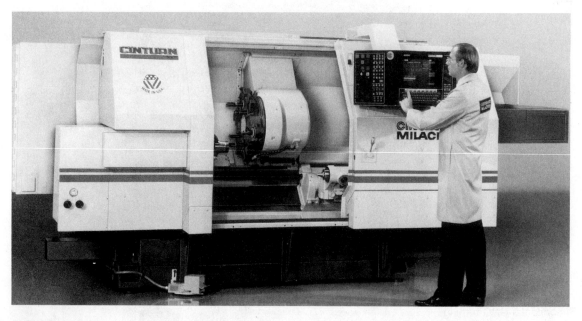

FIGURE 24–4
Many horizontal CNC turning centers are of the slant bed design. (Courtesy of Cincinnati Milacron, Inc.)

A basic CNC turning center, as seen in Figure 24–5, uses only two axes: Z and X. The Z-axis, as mentioned before, travels parallel to the machine spindle. Therefore, a line drawn through the center of the spindle is the Z-axis. For both OD (outside diameter) and ID (inside diameter) operations, a *negative Z (−Z)* is a movement of the saddle toward the headstock. A *positive Z (+Z)* is a movement of

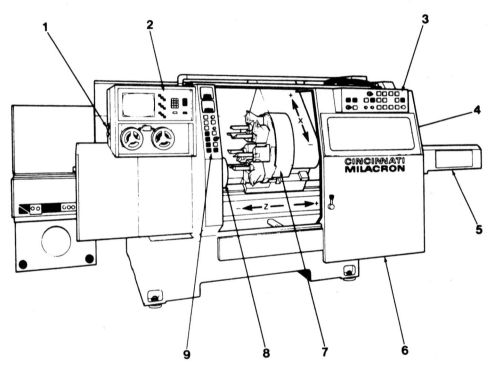

1. HI/LOW CHUCK PRESSURE
2. ACRAMATIC 900TC REMOTE CONSOLE AND TAPE READER
3. ROLLING SHIELD PANEL
4. TAILSTOCK QUILL PRESSURE (IF SUPPLIED) AND AUTO STEADY REST PRESSURE (IF SUPPLIED) ARE MOUNTED ON THE RIGHT SIDE OF BED ROLLING SHIELD IN ILLUSTRATION HIDES THESE CONTROLS.
5. CHIP CONVEYOR
6. ROLLING SHIELD
7. TURRET
8. CHUCK
9. HEADSTOCK PANEL

FIGURE 24–5
Typical CNC turning center with axis directions and major components indicated. (Courtesy of Cincinnati Milacron, Inc.)

the saddle away from the headstock. The X-axis travels perpendicular to the spindle centerline. A *negative X (−X)* moves the cross slide toward the centerline of the spindle and a *positive X (+X)* moves the cross slide away from the spindle centerline.

Regardless of the type of CNC turning center used, a variety of OD and ID operations are performed. These include drilling, turning, facing, boring, and threading. Qualified toolholders must be used to perform OD operations. This means that standard size toolholders are used and the location of the tool insert is held to close tolerances with respect to the rear and opposite sides of the toolholders. This allows holders to be changed without setting gauges. Tool offsets are then dialed in at the control panel. These are used to compensate for tool wear or for minor setup adjustments and will be discussed later in the text.

PROGRAMMING CNC TURNING CENTERS

Most modern CNC turning centers accept the decimal point programming format. Other CNCs accept the word address or interchangeable format with either the EIA (BCD) or ASCII coding. Now, before you begin with a basic CNC turning center program, let's look at some of the typical words used on a CNC turning center. The following list (there are others but these are the important ones to know for now) explains some of the basic words used for a CNC turning center.

- **N** The sequence number is composed of up to four digits preceded by the letter "N" (Nxxxx). This word is used to indicate the block of information which is being processed by the control.

- **G** The preparatory function code is a two-digit number preceded by the letter "G" (Gxx). These codes are used throughout the program to define the various modes of operation.

- **X/Z** Axis dimensions are used to denote the position of the axes. The axes are addressed in decimal point form preceded by the letters "X" or "Z." The sign denotes the direction of travel using the incremental mode, and the position relative to program zero using the absolute mode. (X ± xxx.xxxx).

F   The axis feed rate is controlled by a decimal point number preceded by the letter "F." Feed rates may be programmed in either distance of travel per minute or distance per revolution of the spindle, depending on the selected preparatory function. (Fxxx.x ipm or F.xxxx ipr).

S   Spindle speeds are programmed with a four-digit number preceded by the letter "S." The spindle speed can be programmed in either direct rpm coding or in the constant surface speed (CSS) mode.

T   The turret station and offsets are programmed with a four-digit number preceded by the letter "T." The first and second digits usually identify the turret station, and the third and fourth digits represent the offset. (Txxxx).

M   The miscellaneous function codes are two-digit codes preceded by the letter "M." These codes are used throughout the program to perform functions such as spindle starting and stopping, coolant control, and transmission range selection. (Mxx).

It is extremely important to mention that on any N/C turning center, the turret must be positioned to a location free from interference with the chuck, workpiece, and machine elements before any tool changes are programmed. Failure to comply with this cardinal rule of N/C programming may result in bodily injury and/or machine and tool damage.

> **"Tech Talk"**
>
> Workholding and work-changing equipment, consisting of automatic chuck-changing systems, automatic jaw-changing systems, automated chucks, and automatic pallet-changing systems add considerably to the productivity of CNC turning centers.

Now you can review the following part programs to learn basic CNC turning center programming. Each program provides the drawing of the part to be made, the example program with a line-by-line explanation, and the cut sequence.

The first program (Figure 24–6) is in the decimal point programming format and is broken down into sections (facing, rough turn first diameter, rough turn second diameter, and finish profile) with an illustration for each section so you can easily follow the program and see what is happening. The second program (Figure 24–7) is in the interchangeable variable block format showing the complete program.

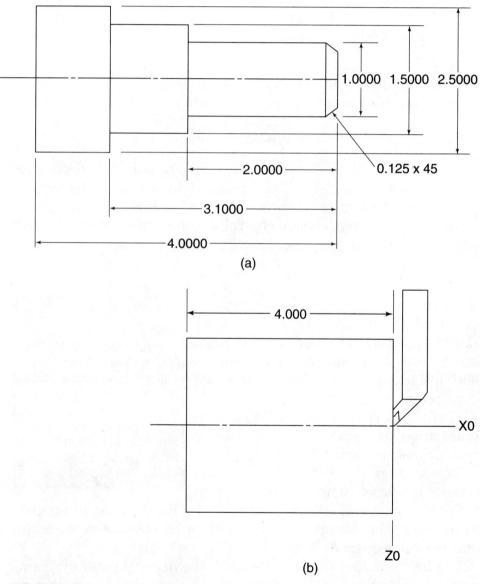

FIGURE 24–6
A simple CNC turning center programming example in the decimal point format.

Program   *Explanation*
G0 T0101   Change to tool 1
G50 S3000 M42   Set spindle limit (S3000), high range (M42)
G96 S250 M3   Set constant surface speed (S250), spindle on
G0 Z.1 X2.7 M8   Rapid to start point 0.100 in. from part, coolant on
G1 Z0 F.015   Feed in to Z0
X −.05   Face part 0.025 in. below center

This is the part of the program to perform the facing cut (program and explanation).

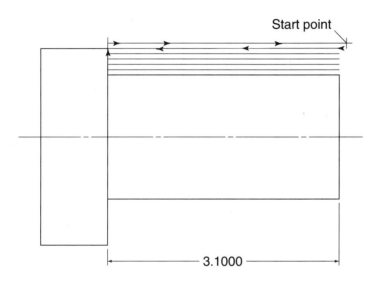

G0 Z.1 X2.7   Rapid to start point
G90 X2.45 Z − 3.085 F.010   Activate canned cycle (G90): OD, ID turning; take cuts of
X2.4   0.100 in. off diameter to a Z depth of −3.085 in.,
X2.3   leaving extra 0.015 in. on diameter and linear
X2.2   dimensions at last step
X2.1
X2.0
X1.9
X1.8
X1.7
X1.6
X1.515

This part of the program performs the rough turning operation using G90 (absolute programming input) and a 0.100 in. depth of cut.

FIGURE 24–6
(continued)

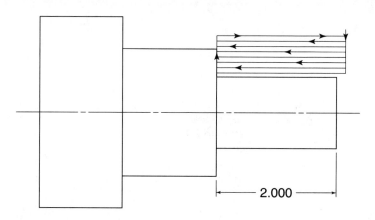

X1.4 Z − 1.985	Continue to take cuts of 0.100 in. on diameter but to a Z distance of − 1.985 in.
X1.3	
X1.2	
X1.1	
X1.015	

This part of the program continues the rough turning operation for the smallest diameter of the part as 0.100 in depth of cut.

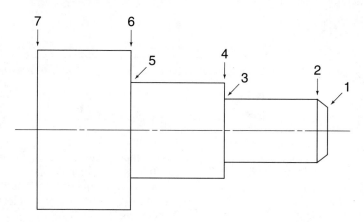

G0 X.75	Rapid to start point 1
G1 Z0 F.010	Feed to Z0 at start point 1
X1.00 Z − .125	Feed to point 2
Z − 2.000	Feed to point 3
X1.5	Feed to point 4
Z − 3.100	Feed to point 5
X2.500	Feed to point 6
Z − 4.000	Feed to point 7
X2.7	Feed clear of part
G0 X6.000 Z6.000	Rapid to a safe distance
M30	

This part of the program finish turns the entire part to size.

FIGURE 24–6
(continued)

SIMPLE CNC LATHE PROGRAMMING • 217

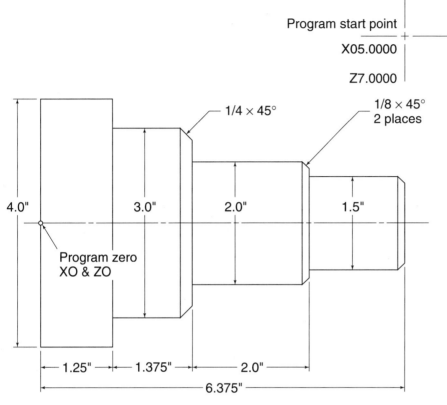

BAR STK. 4.0" × 6.5
T01 (ROUGH FACE AND ROUGH TURN)
T04 FINISH PROFILE (.0468 TNR)

Program				Explanation
010	G97	S300	M41	300 rpm – low-gear range
N20	G90		M13	Absolute positioning, spindle on w/coolant
N30	G70			Inch programming
N40	G00	X50000	Z70000	Rapid to program start pt.
N50	G95	T0101	M06	Inches per rev.—index to tool 1 + assn. offset #1
N60	G96	S400	R50000	Constant surface speed (400 sfm)
N70	G00	X22000	Z63850	Rapid to position to rough face
N80	G01	X-312	F150	Face end of stock
N90			Z65000	Feed away from end of stock

FIGURE 24–7
A CNC turning center programming example in the interchangeable word address format.

Program

				Explanation
N100	G00	X17500		Rapid to pos. for rough turn (.25" depth) leave .010 on shldr. for finish
N110	G01		Z12600	Make 1st rough turn to 1.25 shldr.
N120		X18500	Z13600	Feed + .100 in X + Z (clearance away from part)
N130	G00		Z65750	Rapid out in Z-axis .200 from end of part
N140		X15100		Rapid down to next depth of cut
N150	G01		Z12600	Make second rough turn to 1.25 shldr.
N160		X16100	Z13600	Feed + .100 in X + Z (clearance)
N170	G00		Z65750	Rapid out in Z .200 from end of part
N180		X12500		Rapid down to next depth
N190	G01		Z26350	Make rough turn to 2.625 shldr.
N200		X15100	Z23750	Rough 1/4 × 45° chamfer
N210	G00		Z65750	Rapid out .200 from end of part
N220		X10100		Rapid down to next depth
N230	G01		Z26350	Make rough turn to 2.625 shldr.
N240		X11100	Z27350	Feed + .100 in X + Z (clearance)
N250	G00		Z65750	Rapid out .200 from end of part
N260		X6350		Rapid to next depth
N270	G01		Z63850	Feed in to .100 from end of part
N280		X7600	Z62600	Rough 1/8 chamfer on end of part
N290			Z46350	Feed back to 4.625 shldr.
N295		X8850		Feed up 4.625 shldr.
N300		X10100	Z45100	Rough 1/8 × 45 chamfer on 4.625 shldr.
N310		X12000		Feed up to clear part
N320	G00	X50000	Z70000	Rapid turret back to prog. st. pt.
N330	G97	S1000	M42	Direct rpm (1000) − high-gear range
N340	G90			Absolute positioning
N350	G70			Inch programming
N360	G00	X50000	X70000	Restate current pos. for tool index
N370	G95	T0404	M06	Inches per rev.—index to tool #4 + assn. offset #4
N380	G96	S1000	R50000	Constant surface speed (1000 sfm)
N390	G00	X-470	M13 Z65750	Rapid to ℄ and .200 from end of part
N400	G01	F80	Z63750	Feed in to end of part to begin finish profile
N410		X5976		Feed from ℄ to start of chamfer on end of part
N420		X7500	Z62226	Cuts chamfer on end of part (1/8 × 45°)
N430			Z46250	Feed back in Z to 4.625 shldr.
N440		X8476		Feeds up 4.625 shldr. to start of 2nd 1/8 × 45°C
N450		X10000	Z44726	Cuts 1/8 × 45°C on 4.625 shldr.

FIGURE 24–7
(continued)

N460			Z26250	Feeds across 2.0" dia. to 2.625 shldr.
N470		X12226		Feeds up 2.625 shldr. to start of 1/4 C
N480		X15000	Z23476	Cuts 1/4 × 45°C up 3.0" dia.
N490			Z12500	Feeds across 3.0" dia. to 1.25 shldr.
N500		X21000		Feeds up 1.25" shldr. + clears str. dia. by .100"
N510	G00	X50000	Z70000	Rapid back to st. pt.
N520		T0100	M06	Cancel out active assignable offset
N530		M30		Ends program (M30 shuts off spindle and coolant—also rewinds program)

FIGURE 24-7
(continued)

RECALLING THE FACTS

1. CNC lathes, now called _____, are classified into two types; _____ and _____.
2. The term *OD* stands for _____ and the term *ID* stands for _____.
3. On a CNC turning center, a _____ movement is a movement of the saddle toward the headstock and a _____ movement is a movement of the saddle away from the headstock.
4. On a CNC turning center, _____ movement is a movement of the cross slide toward the spindle centerline, and a _____ is a movement of the cross slide away from the spindle centerline.
5. Rewrite the following decimal point programming block of information in the interchangeable variable length format.

    G01   X1.5   Z − 3.6   F.015   _____

6. The miscellaneous function to reset the program is _____.
7. Write the CNC block in decimal point programming format that will rapid the tool to a Z of 3.500 and an X of 2.750. _____

8. Programming simple sequences.
   a.

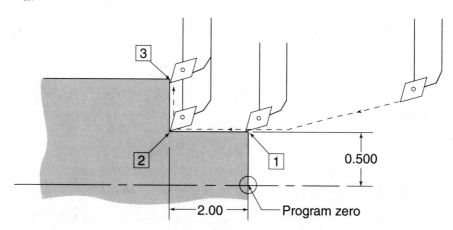

Fill in the missing G – X and Z values for this cut sequence.

Rapid to position 1	N100	G0	X5.	Z.1		
Feed to position 2	N110	G95	G	X	Z	F.007
Feed to position 3	N120	U.4				

Note: An address of U may be used to specify incremental X axis moves and an address of W may be used to specify incremental Z axis moves.

b.

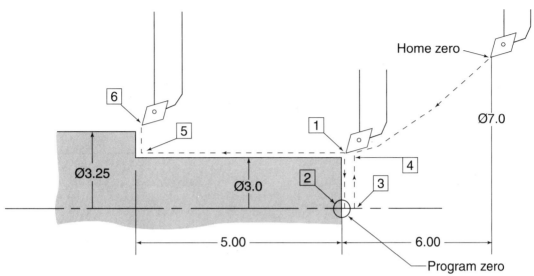

Fill in the missing G – X and Z values for this cut sequence.

Set home position	N110	G50	X7.	Z6.		
Rapid to position 1	N115	G___	X3.5	Z___	G96 S400	M3
Face part to position 2	N120	G___	X – .032	F.013		
Rapid away to position 3	N125	G0	W.1			
Rapid to position 4	N130	X3.	Z.1			
Turn 3.0 dia. to position 5	N135	G___	Z___			
Incremental move to face shoulder/clear part at position 6	N140	U.2				
Rapid to home zero position	N145	G___	X___	Z___	M5	
Program end	N150	M30				

UNIT 25

Simple CNC Machining Center Programming

KEY TERMS

vertical machining centers
horizontal machining centers

TYPES OF CNC MACHINING CENTERS

Known in the 1960s as ATCs, or automatic tool changers, machining centers originated out of their capability to perform a variety of machining operations on a part by changing their own cutting tools. In many cases, the purchase of a machining center served as a shop's introduction to numerical control. Today, machining centers have become well-accepted as a separate class of machine tool and their usage continues to expand from stand-alone job shop applications to flexible manufacturing cells and systems. However, the increased cutting time of the machine places considerable wear and tear on this relatively new breed of machine tools.

Machining centers, just like turning centers, are classified as either vertical or horizontal. *Vertical machining centers* (Figure 25–1) continue to be widely accepted and used, primarily for flat parts and where three-axis machining is required on a single part face, such as in mold and die work. *Horizontal machining centers* (Figure 25–2) are also widely accepted and used, particularly because parts can be machined on multiple sides in one clamping of the workpiece and because they lend themselves to easy and accessible pallet transfer when used in a cell or system application. The ability to perform a number of different operations, such as milling, drilling, boring, counterboring, and tapping, in a single workpiece setup while changing its own cutting tools is what has made the horizontal machining center such a valuable, productive tool in a manufacturing shop.

SIMPLE CNC MACHINING CENTER PROGRAMMING • 223

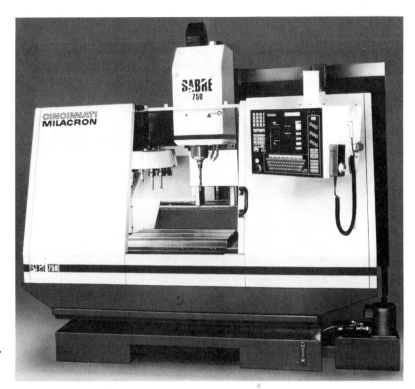

FIGURE 25–1
Vertical machining center.
(Courtesy of Cincinnati
Milacron, Inc.)

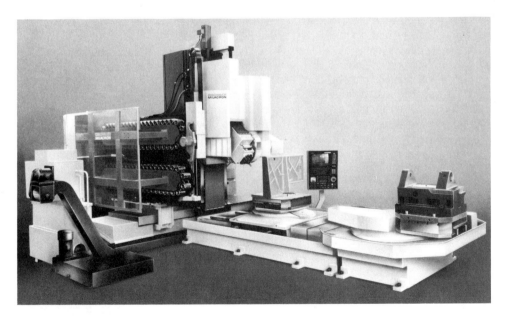

FIGURE 25–2
Horizontal machining center. (Courtesy of Cincinnati Milacron, Inc.)

> **"Tech Talk"**
>
> Five N/C programmers will usually generate five different programs when given the exact same assignment. However, one of these programs will generally be the best. That program will be the one that makes the most efficient and effective use of the machine tool, the cutting tools, the fixtures, and the N/C programming/coding techniques available to the programmer.

Typical motions of machining centers include the primary axes of X, Y, and Z. However, horizontal machining centers (Figure 25–3) have a rotary index table which allows the part to be indexed from one side to another. Work tables and the B-axis will be discussed in more detail later in the text. Selection of either a vertical or a horizontal machining center mainly depends on the part type, size, weight, and application of the work to be done. Each has its own specific advantages and disadvantages.

General advantages and disadvantages of vertical and horizontal machining centers

Vertical machining centers		Horizontal machining centers	
Advantages	**Disadvantages**	**Advantages**	**Disadvantages**
• Generally less costly. • Thrust is absorbed directly into machine table during deep tool thrust operations such as drilling. • Ideal for large flat plate work, moldwork, and other single-surface three-axis contouring. • Heavy tools can be used without much concern about deflection.	• Not suitable for large, boxy, heavy parts. • As workpiece size increases, it becomes more difficult to conveniently look down into the cut. • Extensive chip buildup obstructs view of the cut and recuts chips. • On large verticals, head weights and distance from the column can cause head drop, loss of accuracy, and chatter.	• More flexible overall. • Table indexing capability enables multiple sides of a workpiece to be machined in one setting and clamping. • Chips drop out of the way during machining, providing an uncluttered view of the cut and preventing recutting of chips. • Operator's station is to one side of the column, providing good line-of-sight control. • Pallet shuttle exchange mechanisms are open, accessible, and easy to service. • Ideally suited for large, boxy, heavy parts.	• Generally more costly. • Difficult to load and unload large, flat, plate-type parts. • High thrust must be absorbed by tombstones, fixtures, or right-angle braces. • Heavy tools can deflect.

FIGURE 25–3
General advantages and disadvantages of vertical and horizontal machining centers.

Machining center innovations and developments have brought about the following improvements:

- improved flexibility and reliability
- increased feeds, speeds, and overall machine construction and rigidity
- reduced loading, tool-changing, and other noncutting time
- greater MCU capability and compatibility with systems
- reduced operator involvement
- improved safety features and less noise

PROGRAMMING CNC MACHINING CENTERS

Now, before you begin with a basic CNC machining center program, let's look at some of the typical words used for a CNC machining center (there are others but these are the basic ones to know for now).

- **N** Sequence number coding, introduced by N. It is a five-character code; one letter and four numerals (Oxxxx or Nxxxx). It is used to identify a block of information. It is informational rather than functional.

- **G** Preparatory function coding, introduced by G. It is a three-character code; one letter and two numerals (Gxx). It is used for control of the machine. It is a command that determines the mode of operation of the system.

- **X** X-axis coordinate information code, introduced by X. It may contain up to nine characters; one letter, one sign, and up to seven numerals (Xxxxx.xxx millimeters). This word is used to control the direction of table travel and position.

- **Y** Y-axis coordinate information coding, introduced by Y. This is identical to the X word, only using the Y address.

- **Z** Z-axis coordinate information coding, introduced by Z. This is identical to the X word, only using the Z address.

- **R** Z-axis coordinate information coding, introduced by R. It may contain up to nine characters; one letter, one sign, and up to seven numerals (Rxxx.xxxx inches, Rxxxx.xxx millimeters). It is used to control the positions of the Z slide at rapid traverse during positioning.

F   Feed rate coding for X-, Y-, and/or Z-axes, introduced by F. It may contain up to five characters in the code; one letter and up to four numerals (Fxxx.x inches/minute, Fxxxx millimeters/minute). It is used for controlling the rate of longitudinal, vertical, and cross travel.

S   Spindle-speed coding for rate of rotation of the cutting tool, introduced by S. It may contain up to five characters in the code; one letter and up to four numerals (Sxxxx). This is the actual rpm desired for cutting.

T   Tool number coding, introduced by T. It is a nine-character code; one letter and eight numerals (Txxxxxxxx). It determines the next tool to be used.

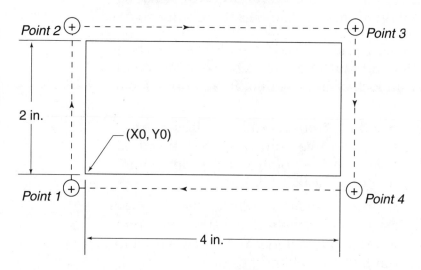

Program	Explanation
N001 G0 X − .25 Y − .25 M03	Rapid to Point 1 allowing for cutter radius, spindle on CLW
N002 G01 Z − .250 F20.	Feed down in Z .250 at 20 inches per minute to depth
N003 Y 2.25	Feed to Point 2 at 20 inches per minute
N004 X 4.25	Feed to Point 3 at 20 inches per minute
N005 Y − .25	Feed to Point 4 at 20 inches per minute
N006 X − .25	Feed to Point 1 at 20 inches per minute
N007 G0 Z 5.0 M30	Rapid retract cutter 5 inches above part and reset program to sequence number N001

FIGURE 25–4
Example of decimal point programming.

M Miscellaneous function coding, introduced by M. It is a three-character code; one letter and two numerals (Mxx). It is used for various discrete machine functions.

Earlier in the text, you studied some "canned cycles" and what they do, as well as some standard G codes. You may want to go back and review these so you have a better understanding of what is happening on the machine as you study the example programs.

Now you can review the following two simple part programs to learn basic CNC machining center programming. Each program provides the drawing of the part to be made, the example program with a line-by-line explanation, and the cut sequence. These examples are in the decimal point programming format and illustrated in Figures 25–4, and 25–5.

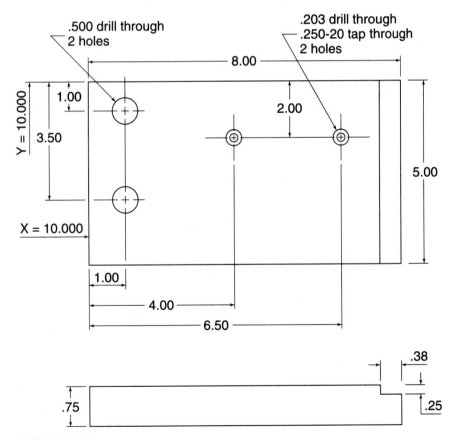

FIGURE 25–5
Example of decimal point programming.

In this example, the X and Y setup/alignment positions are set at X = 10.000 from the left edge of the part and Y = 10.000 from the top edge of the part. Every dimension in the X direction in this example will be added to X of 10.000. (This is because the dimensions in the X direction are from the left edge of the part.) Every dimension in the Y direction in this example will be subtracted from Y of 10.000. (This is because Y of 10.000 is set at the top of the part in this case and the dimensions in the Y direction are from the top edge of the part.) Remember the Cartesian Coordinate System and the four quadrants? What quadrant will this part be programmed in?

Program	Explanation
N001 G54 G90 G80 G40 G00	Absolute input, cancel, rapid
N002 M6 T1	Tool change to tool 1 (center drill)
N003 M03 S1500	Turn on spindle, 1500 RPM
N004 X11. Y9.	Rapid to first .500 hole
N005 G81 G99 X11. Y9. Z − .2 R.05 F3.	Canned cycle, center drill first .500 hole, rapid to .050 above part feed − .200 into part at 3.0 IPM.
N006 Y6.5	Center drill second .500 hole.
N007 X14. Y8.	Center drill first .250 − 20 hole
N008 X16.5	Center drill second .250 − 20 hole
N009 G80	Cancel drill cycle G81
N010 G00 Z10. M5	Rapid to tool change position and turn off spindle
N011 M6 T2	Tool change to tool 2 (.203 drill)
N012 M3 S850	Turn on spindle, 850 RPM
N013 G81 G99 X16.5 Y8. Z − .850 R.05 F3.	Canned cycle drill first .203 hole Rapid to .050 above part, feed .810 (enough to break through bottom of part), at 3.0 IPM
N014 X14.	Drill second .203 hole
N015 G80	Cancel drill cycle G81
N016 G0 Z10. M5	Rapid to tool change position and turn off spindle
N017 M6 T3	Tool change to tool 3 (.250 − 20 TAP)
N018 M3 S400	Turn on spindle, 400 RPM
N019 G84 X14. Y8. Z − .925 R.05 F20.	Tap first .250 − 20 hole, rapid to .050 above part, feed .900 (enough to get full thread through bottom of hole). At 20.0 IPM
N020 X16.5	Tap second .250 − 20 hole
N021 G80	Cancel G84 tap cycle
N022 G0 Z10. M5	Rapid to tool change position and turn off spindle
M6 T4	Tool change to tool 4 (.500 drill)
M3 S320	Turn on spindle, 320 RPM
G81 G99 X11. Y9. Z − .925 R.05 F1.5	Canned cycle drill first .500 drill Rapid to .050 above part, feed .925

FIGURE 25–5
(continued)

Y6.5	(enough to break through bottom of part at 1.5 IPM).
G80	Drill second .500 drill
G0 Z10. M5	Cancel drill cycle G81
	Rapid to tool change position and turn off spindle
M6 T5	Tool change to tool 5 (.750 end mill)
M3 S400	Turn on spindle 400 RPM
G0 17.995 Y10.475 R.05	Rapid to starting X and Y position
G01 Z − .25 F2.5	Feed to depth in Z at 2.5 IPM
Y4.625	Feed in Y axis to top of part
G0 Z10. M5	Rapid to tool change position
M30	Reset program

Calculations for X and Y Dimensions

X Start position

  10.000 — X align position
+ 8.000 — Length of part
  18.000
− .380 — Edge of cut surface from end of part
  17.620
+ .375 — Add half of .750 tool diameter.
  17.995 — X coordinate value

Y Start position

  10.000 — Y align position
+ .375 — Add half of .750 tool diameter
  10.375
+ .100 — Add additional clearance
  10.475 — Y Coordinate value

Y End position

  10.000 — Y align position
−5.000 — Width of part
  5.000
− .375 — Subtract half of tool diameter.
  4.625 — End Y coordinate value

FIGURE 25-5
(continued)

It is important to remember as you begin your review of these programs that there is a great deal more involved in programming machining centers than the brief overview presented here. Items such as tool assemblies, setup information, and tool compensation, as well as work tables, alignment, and other items play an important part in the success of any CNC machining center installation. However, for the sake of concentrating on the basics of programming, these items have been omitted. You will learn more about some of these in the next unit and in your further studies of CNC programming.

RECALLING THE FACTS

1. _____ machining centers are primarily used for flat parts and where three-axis machining is required on a single part face, such as in mold and die work.

2. _____ machining centers are normally used where parts can be machined on multiple sides in one clamping of the workpiece.

3. Typical motions of machining centers include the primary axes of _____, _____, and _____.

4. One advantage of vertical machining centers is that they usually _____.

5. One advantage of horizontal machining centers is that they usually _____.

6. CNCs for machining centers accept either the interchangeable word address format or the _____ format.

7. Rewrite the following block of information in the decimal point programming format.

    G81   X + 100000   Y − 5750   R10000   Z − 8750   M03

8. Write the CNC block in decimal point programming to rapid to an X of 5.000, a Y of 2.250, and a Z of 1.000.

9. Write the CNC block in decimal point programming to feed to an X of 4.500, a Y of −1.437, a Z of −1.200, a feed of 5.5 inches per minute, a spindle speed of 1500 rpm, and the spindle on CLW with coolant.

10. Using the tapping canned cycle and a feed rate of 12 inches per minute, write the CNC blocks in decimal point programming format to tap the following hole locations:

    X of 8.300 and Y of 5.625 _____
    X of 8.300 and Y of 5.000 _____
    X of 8.300 and Y of 4.375 _____
    X of 9.250 and Y of 4.375 _____

X of 9.250 and Y of 5.000 _____
X of 9.250 and Y of 5.625 _____

11. Programming simple sequences.

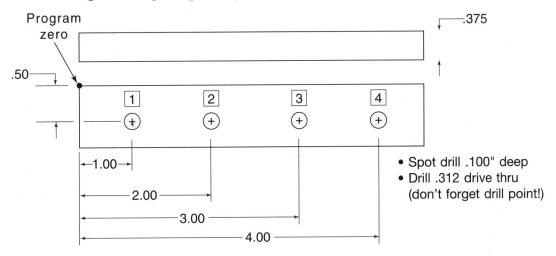

- Spot drill .100" deep
- Drill .312 drive thru
  (don't forget drill point!)

Fill in the missing G – X – Y and Z values for this cut sequence.

Description	N					
Tool change to tool 1	N10	T1	M06			
Rapid to position 1	N20	G☐	X☐	Y☐	M3	S1500
Spot drill position 1	N30	G☐	R.1	Z☐	F3.	
Spot drill position 2	N40	X☐				
Spot drill position 3	N50	X☐				
Spot drill position 4	N60	X☐				
Cancel drill cycle	N70	G80				
Rapid to tool change position —Turn off spindle	N80	G00	Z10.	M5		
Tool change to tool 2	N90	M6	T2			
Rapid to position 4	N100	G☐	X☐	Y☐	M3	S900
Drill to depth position 4	N110	G☐	R.1	Z☐	F3.	
Drill position 3	N120	X☐				
Drill position 2	N130	X☐				
Drill position 1	N140	X☐				
Cancel drill cycle	N150	G☐				
Rapid to tool change position Turn off spindle	N160	G☐	Z10.	M5		

232 • UNIT 25

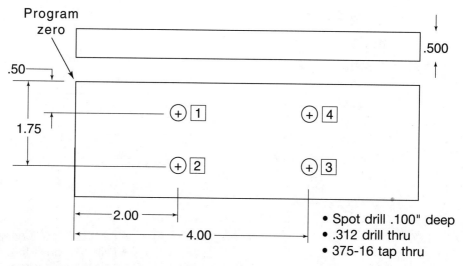

- Spot drill .100" deep
- .312 drill thru
- 375-16 tap thru

Fill in the missing G – X – Y and Z values for this cut sequence

Description	N					
Tool change to tool 1	N10	T1	M06			
Rapid to position 1	N20	G00	X2.0	Y-.5	M3	S1500
Spot drill position 1	N30	G81	R.1	Z-.1	F3.	
Spot drill position 2	N40	Y-1.75				
Spot drill position 3	N50	X4.0				
Spot drill position 4	N60	Y-.5				
Cancel drill cycle	N70	G80				
Rapid to tool change position						
Turn off spindle	N80	G00	Z10.	M5		
Tool change to tool 2	N90	M6	T2			
Rapid to position 4	N100	G00	X4.0	Y-.5	M3	S1000
Drill to depth position 4	N110	G81	R.1	Z-.6	F3.	
Drill position 3	N120	Y-1.75				
Drill position 2	N130	X2.0				
Drill position 1	N140	Y-.5				
Cancel drill cycle	N150	G80				
Rapid to tool change position						
Turn off spindle	N160	G00	Z10.	M5		
Tool change to tool 3	N170	M6	T2			
Rapid to position 1	N180	G00	X2.0	Y-.5	M3	S500
Tap position 1	N190	G84	R.1	Z-.6	F31.	
Tap position 2	N200	Y-1.75				
Tap position 3	N210	X4.0				
Tap position 4	N220	Y-.5				
Cancel tap cycle	N230	G80				
Rapid to tool change						
Position – turn off spindle	N240	G00	Z10.	M5		

UNIT 26

Some Additional Programming Functions

KEY TERMS

circular interpolation	B-axis	G03
J word	G02	G41
cutter diameter compensation	tool offsets	I word
rotary index tables	G40	G42
K word	probing	stylus

In this unit you will touch on some of the more widely-used functions that programmers and operators must know. Each is used on a variety of different CNC machines and in different ways. At this time it is only important that you have a basic understanding of what each function is, what it does, and generally how it works.

CIRCULAR INTERPOLATION

Many applications in CNC machining require the cutting tool to move in a circular path. *Circular interpolation* allows the programmer to move the cutting tool in a circular path that can range from a small arc segment to a full circle (360 degrees). The cutter path along the arc is generated by the control system. On some CNCs, cutting a full 360 degrees can be accomplished in one block of information. Older controllers required four arcs of 90 degrees each to be programmed. The total number of blocks programmed will normally vary depending on the specific type of circular interpolation used.

In addition to the feed rate, there are four basic elements necessary to program circular interpolation:

- preparatory function (G02) or (G03)
- start point

- center point
- end point

The preparatory function codes *G02* and *G03* are used for programming circular interpolation. These codes determine the direction of the circular path. G02 is the command to begin circular interpolation clockwise (CW), and G03 is the command to begin circular interpolation counterclockwise (CCW). These codes are programmed in the block where circular interpolation should begin. G02 and G03 are modal codes, which means, as you learned earlier, that they stay in effect until a new preparatory function (G code) is programmed.

The start point is usually the end point of a previous arc (circular interplation) or the end point of a line (linear interpolation). The start point is always described by X, Y, and/or Z words and it positions the cutter for the following circular move.

The center point is the center of the circular arc. The center point is described by I, J, and K words. The *I word* describes the X-coordinate value, the *J word* describes the Y-coordinate value, and the *K word* describes the Z-coordinate value. If the control knows the start point, it knows the radius. On some controllers, the radius of the arc can take the place of the center point. Therefore, on some controls, two methods of programming circular interpolation can be used; the center point method and the radius method.

The end point is the end of the circular move or the point on the arc where circular interpolation stops. The end point can be described by X, Y, and/or Z words or it can be specified as the number of degrees through which to rotate.

During programming, it is extremely important to know precisely how circular interpolation is handled because mistakes can be very dangerous and costly, as you have already learned. Always study the specific operator/programmer's manual before attempting to program any CNC machine tool.

Figures 26–1 and 26–2 show two methods of programming circular interpolation for each example; the center point method and the radius method. Each method provides an explanation as to what is happening on each line of the program.

SOME ADDITIONAL PROGRAMMING FUNCTIONS • 235

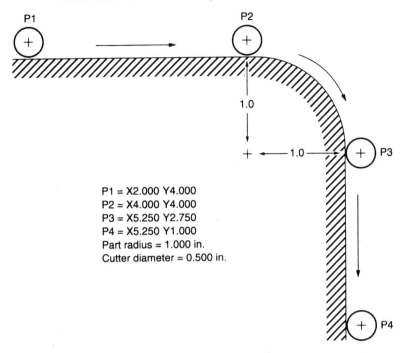

Clockwise Circular Interpolation for a 90° Arc

P1 = X2.000 Y4.000
P2 = X4.000 Y4.000
P3 = X5.250 Y2.750
P4 = X5.250 Y1.000
Part radius = 1.000 in.
Cutter diameter = 0.500 in.

CENTER POINT METHOD

N20	G01	X2.	Y4.	F6.	Move to P1 at 6 IPM
N21	X4.				Move to P2 (start point)
N22	G02	X5.25	Y2.75	I4. J2.75	C.I from P2 to P3 (end point) CW
N23	G01	Y1.			Move to P4

RADIUS METHOD

N20	G01	X2.	Y4.	F6.	Move to P1 at 6 IPM
N21	X4.				Move to P2 (start point)
N22	G02	X5.25	Y2.75	R1.25	C.I. from P2 to P3 (end point) CW
N23	G01	Y1.			Move to P4

Figure 26–1
First example of circular interpolation showing both center point and radius methods.

Counterclockwise Circular Interpolation for a 180° Arc

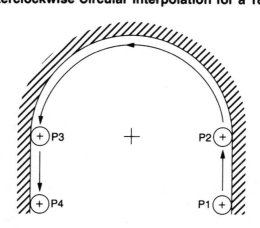

P1 = X1.75 Y−1.5
P2 = X1.75 Y0
P3 = X−1.75 Y0
P4 = X−1.75 Y−1.5
Part radius = 2.000 in.
Cutter diameter = 0.500 in.
Arc center = 0.0

CENTER POINT METHOD

```
N30  G01  X1.75   Y−1.5  F6.        Move to P1 at 6 IPM
N31       Y0                         Move to P2 (start point)
N32  G03  X−1.75  Y0  I0  J0         C.I. from P2 to P3 (end point) CCW
N33  G01  Y−1.5                      Move to P4
```

RADIUS METHOD

```
N30  G01  X1.75   Y−1.5  F6.         Move to P1 at 6 IPM
N31       Y0                          Move to P2 (start point)
N32  G03  X−1.75  Y0  R1.75           C.I. from P2 to P3 (end point) CCW
N33  G01  Y−1.5                       Move to P4
```

Figure 26–2
Second example of circular interpolation showing both center point and radius methods.

TOOL OFFSETS

A *tool offset* is a value that the CNC control reads and stores as the radius or length of a cutting tool. Offset values are manually entered in the MCU to make adjustments to the cutting tool path based on the length and radius of the specific cutter being used. Offsets define the cutter geometry and position for the CNC's computer to use in determining the cutter path. Tool length offsets and tool radius offsets are not part of the actual program; they are stored in the control in tool offset registers. Essentially, each tool pocket or tool location has length and radius offset registers.

Offsets are normally entered during the setup and can be manually changed to adjust for available cutter sizes or to suit the new set of cutting conditions. When a cutter is changed, the offset for that tool is reentered to suit the new cutter and cutting conditions.

Offsets have four basic uses:

1. They allow a "close-to" diameter cutter to be used when the originally intended cutter diameter is unavailable.
2. Entering an offset value that doesn't match the cutter being used allows you to "trick" the control into cutting the part either undersize or oversize (see Figure 26–3). This can be very

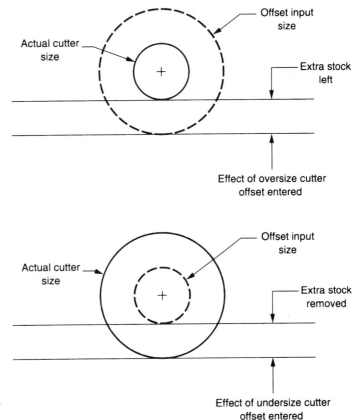

Figure 26–3
Entering offset values for oversize or undersize cutter diameters allows extra stock to be left or removed.

useful to leave finish stock or to recut the part to remove more material without changing the program.

3. They can be used to compensate for tool wear. As a tool dulls, the offset can be changed to "trick" the control into moving the cutter slightly closer to the cutting surface. However, this is not a good practice because surface finish and other problems may arise from using dull tools. Dull tools should be replaced with sharp cutting tools.

4. Through the use of two different sets of tool offsets, rough and finish passes may be made using the same program and the same cutting tool. The first offset tells the control that the cutter is larger than it really is, which causes the control to back the cutter away from the part, thereby leaving finish stock. Recutting the part using the second offset—which represents the correct cutter diameter—will then result in a part with the correct finish size.

CUTTER DIAMETER COMPENSATION

Cutter diameter compensation (CDC) allows the cutter diameter to be changed similar to tool offsets by telling the control that the machine has a cutter that is larger or smaller in diameter than it actually is. Cutter diameter compensation adds additional capability to the machine and control system by:

- controlling the size of milled close-tolerance slots, pockets, bores, and steps, and
- using a cutter of a different diameter than that originally intended when the part is programmed.

CDC permits oversize and undersize cutters to be used while still maintaining the programmed part geometry. The difference between the cutter diameter programmed and the one used—known as a *CDC value*—generally ranges from −1.0000 in. to +1.0000 in. Methods for inputting and compensating for cutter size differences vary among control manufacturers.

When cutter diameter compensation is being used, a CDC value can be entered into the control for each tool number programmed. Inputting a CDC value for one tool does not affect CDC values for other tools. The CDC value becomes active when the appropriate tool is loaded into the spindle. If the value is changed after the tool is located, it becomes active on the next span prepared by the control system. CDC, effective only when the appropriate codes are programmed, can be programmed for both linear and circular interpolation.

When programming a part for any milling operation, the operator/programmer must take three main factors into consideration:

1. The coordinates/dimensions of the surfaces to be cut.
2. The cutter direction and position relative to the surface or programmed line to be machined. (When CDC is not in effect, the cutter's position is already fixed by coordinate locations in the program.)
3. The radius of the cutting tool.

If CDC will be used, specific codes must be added. The G codes used for programming CDC are:

- *G40*—cancel cutter compensation code for G41 or G42.
- *G41*—cutter diameter compensation left of line. This code directs the cutter to the left side of the programmed cutting line (looking in the direction of the cutter path).
- *G42*—cutter compensation right of line. This code directs the cutter to the right side of the programmed cutting line (looking in the direction of the cutter path).

When deciding which code to use (G41 or G42), the operator/programmer must consider the direction of the cutter path and the proper cutter position. After determining the direction of the cutter movement, he or she must visualize "walking directly behind" the cutter. This determines the cutter position relative to the drive surface or cutting line. Figures 26–4 and 26–5 illustrate how cutter diameter compensation is used on many CNC controls today.

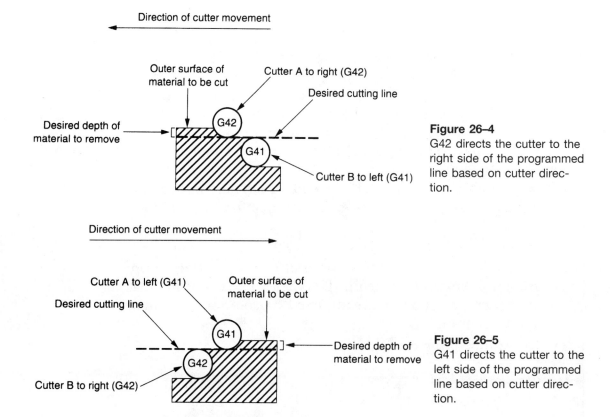

Figure 26–4
G42 directs the cutter to the right side of the programmed line based on cutter direction.

Figure 26–5
G41 directs the cutter to the left side of the programmed line based on cutter direction.

ROTARY INDEX TABLES

Rotary index tables primarily provide versatility for horizontal machining centers, as you have already learned to some extent. With proper fixturing and clamping of the workpiece, the entire part can be machined in one setup (see Figure 26–6). Rotary table motion is designated by the *B-axis* which must be aligned the same as the X-, Y- or Z-axes. An example of a rotary index table showing the orientation of all axes is shown in Figure 26–7.

Rotary tables are bidirectional (will rotate in two directions) and will index using the shortest path to get to the end position. Work tables on horizontal machining centers can index to a position based on how the B-axis is programmed. This means they can index and stop at every degree for some machining work (remember, a complete circle contains 360 degrees), every half of a degree (720 positions), or positions even finer than half of a degree. Other rotary tables are used to provide feed rate control of the B-axis for machining.

SOME ADDITIONAL PROGRAMMING FUNCTIONS • 241

> **"Tech Talk"**
>
> Company inventory consists of raw materials, work-in-process, and finished goods. Raw materials are purchased items such as castings, bar stock, chemicals, etc. used to make products. Work-in-process consists of items in stages of completion, i.e., parts waiting to be machined or being machined. Finished goods are finished products, assemblies, or subassemblies in stock, on a loading dock, or in a warehouse.

Programming a machine tool with a B-axis rotary index table takes practice just like programming any other axis on any machine tool. However, with any type of programmable index table, programmers and operators should always make sure that the cutting tool is retracted and located in a safe position so that the table will index without hitting the cutting tool.

PROBING

The probe (sometimes called a *touch-trigger* or *surface-sensing probe*) is a very sensitive measuring device used for a variety of applications. **Probing** is used to

Figure 26–6
Conventional index table with part mounted on machine table. (Courtesy of Monarch Machine Tool Company.)

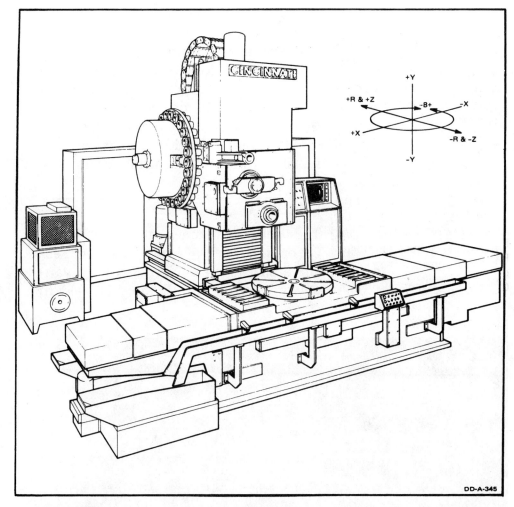

Figure 26–7
Rotary index table and machine with axis orientation shown. (Courtesy of Cincinnati Milacron, Inc.)

trigger electronically the programmed surface reference in X, Y, or Z by touching a surface and then making automatic compensation for the measured values through direct feedback to the control. Probing is used on CNC turning centers, and machining centers, as seen in Figures 26–8 and 26–9.

The probe itself is a high-precision switch which can be loaded and automatically selected from the tool storage matrix of a machining center, as seen in Figure 26–10. The probe *stylus* is attached

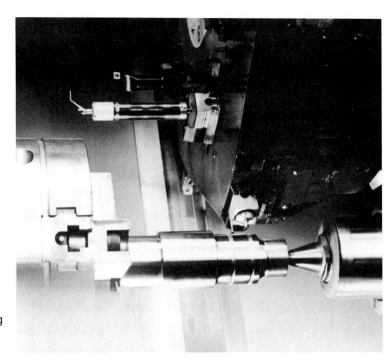

Figure 26–8
Probing on a CNC turning center. (Courtesy of Cincinnati Milacron, Inc.)

Figure 26–9
Probing on a CNC machining center. (Courtesy of Cincinnati Milacron, Inc.)

Figure 26–10
The surface sensing probe loaded in the tool storage matrix of a horizontal machining center. (Courtesy of Cincinnati Milacron, Inc.)

to the switch, and when the stylus moves and comes in contact with a surface, it is deflected, and the probe captures the current values in the machine's active position register. The values from the active position register can then be used to locate the setup point, establish tool clearance planes, or for on-machine part inspection following completion of all machining operations.

The surface-sensing probe contains three parts:

1. the probe body with an interchangeable stylus (the stylus makes physical contact with the part),
2. the noncontacting modules (one module is mounted on the nose of the machine spindle, whereas another is mounted on the probe itself), and
3. the control and the interface printed circuit board.

Programming for the probe takes extra time and requires extra skill over and above the machining skills already discussed. This is because of all the extra thinking and planning the programmer must do and all the extra programming statements and logic that must be added into the program to move the probe to the various inspection and touch points.

RECALLING THE FACTS

1. _____ allows the programmer to move the cutter in a circular path ranging from a small arc to a full circle (360 degrees).
2. _____ is the command to begin circular interpolation clockwise while _____ is the command for circular interpolation counterclockwise.
3. The four elements necessary to begin circular interpolation using the _____ method are:
   a. _____
   b. _____
   c. _____
   d. _____
4. Two methods of programming circular interpolation are the _____ method and the _____ method.
5. An _____ is the value that the control reads and stores as a radius or length of the cutting tool; it defines the cutter _____ and _____ for the CNC's computer to use in determining the cutter path.
6. Each tool pocket or tool location has _____ and _____ offset registers.
7. _____ offsets have four basic uses:
   a. They allow a _____ diameter to be used when the programmed cutter is not available.
   b. Entering an offset value allows you to "trick" the control into cutting the part either _____ or _____.
   c. They can be used to compensate for _____ _____.
   d. They allow for _____ and _____ passes to be made with the same program and the same cutting tool.
8. _____ allows the cutter diameter to be changed similar to tool offsets by telling the control that the machine has a cutter that is larger or smaller than it actually is.
9. _____ permits _____ and _____ cutters to be used while still maintaining the programmed part geometry.

10. In _____, the _____ code directs the cutter to the left side of the programmed cutting line, the _____ code directs the cutter to the right side of the programmed cutting line, and the _____ cancels both the _____ and _____ code.

11. With proper fixturing and clamping of the part, _____ _____ _____ on horizontal machining centers permit the entire part to be machined in one setup.

12. _____ table motion on horizontal machining centers is applied through the _____-axis.

13. The _____, sometimes called a *touch* _____ or *surface* _____ is a high-precision measuring device used to touch a surface and then make automatic compensation for the measured values through _____ _____ to the control.

14. The _____ _____ is attached to the switch and when the _____ moves and comes in contact with a surface, it is deflected and the probe captures the current values in the machine's active machine _____.

15. The surface-sensing probe contains three parts:
    a. _____,
    b. the noncontacting _____ (one is mounted on the machine spindle and the other is mounted on the _____), and
    c. the control and the interface _____.

SECTION 4
REVIEW

KEY POINTS

- Engineering creates product designs and detail drawings for manufacturing and assembly.
- Manufacturing engineering plans how parts are to be manufactured.
- Inventory control determines how many lots or batches of parts are to be dispatched to the shop at one time and what the schedule or priority will be for those parts to be manufactured.
- The process plan is a description of what needs to be done to complete the manufacturing of the part.
- Tool design designs fixtures and tracks the fixture's manufacture, assembly, and inspection.
- Setup instructions are the written or graphic instructions from the programmer to the operator describing how the part should be set up on the CNC machine, what the alignment or program zero or reference positions/points are, and how to hold and orient the part at the machine.
- A revision is a change to an engineering design, process plan, or N/C program. Revision numbers should always agree

I'm starting to really understand CNC programming! But let's review anyway.

among the engineering drawing, the process plan, and the N/C program.

- Check-out consists of checking the output manuscript, line by line, for any glaring mistakes, problems, or omissions.
- Prove-out consists of the programmer being out in the shop at the machine with the operator checking out a new program the first time it is run on the machine.
- The amount of time spent on a prove-out will depend on four things: the length of the program, the care taken during the preparation of the program, the care taken during the setup, and the care taken during check-out.
- A prove-out should always be done to any new program before running continuous parts.
- Prove-outs can also consist of plots (a printed outline of the cutter path on paper indicating rapid and feed cutter path moves relative to the part surfaces), a dry run (running through the program with tools and the fixture loaded, for example, but no part), and running the program in single block mode (a mode which lets the program execute only one block at a time).
- Always plan a safe prove-out and production run and stay alert and focused on the task at hand.
- CNC lathes, now commonly called *turning centers*, are classified as either vertical or horizontal.
- OD stands for outside diameter and ID stands for inside diameter.
- On a CNC turning center, a negative Z ($-Z$) is a movement of the saddle toward the headstock, a positive Z ($+Z$) is a movement of the saddle away from the headstock, a negative X ($-X$) is a movement of the cross slide toward the spindle centerline, and a positive X ($+X$) is a movement of the cross slide away from the spindle centerline.
- Vertical machining centers are used primarily for flat parts and where three-axis machining is required on a single part face, such as in mold and die work.

- Horizontal machining centers are primarily used where parts can be machined on multiple sides in one clamping of the workpiece.
- Circular interpolation allows the programmer to move the cutting tool in a circular path that can range from a small arc segment to a full circle (360 degrees).
- In addition to the feed rate, there are four basic elements necessary to program circular interpolation: (1) preparatory function (G02—clockwise rotation or G03—counterclockwise rotation), (2) start point, (3) center point, and (4) end point.
- On some controllers, two methods of programming circular interpolation can be used, the center point method and the radius method.
- The center point location in circular interpolation is described using the I word (X-coordinate value), J word (Y-coordinate value), and K word (Z-coordinate value).
- A tool offset is a value that the CNC control reads and stores as the radius or length of a cutting tool.
- Cutter diameter compensation (CDC) allows the cutter diameter to be changed similar to tool offsets by telling the control that the machine has a cutter that is larger or smaller in diameter than it actually is.
- G codes used in cutter diameter compensation are G40 (cancel CDC), G41 (CDC left of line), and G42 (CDC right of line).
- Rotary index tables provide rotary table motion to horizontal machining centers through the table's B-axis which is bidirectional (will rotate in two directions).
- Probing, sometimes called *touch-trigger* or *surface-sensing probe*, is used to trigger electronically the programmed surface reference in X, Y, or Z by touching a surface and then making automatic compensation for the measured values through direct feedback to the control.
- The probe stylus is attached to the switch such that when the stylus moves and comes in contact with a surface, it is deflected, and the probe captures the current values in the machine's active position register.

REVIEW QUESTIONS AND EXERCISES

1. Describe the difference between engineering design and manufacturing engineering.
2. Tool design is involved in creating process plans. (True/False)
3. In your own words, describe what setup instructions are and why they are important to the operator.
4. As a programmer or operator, why is it important to always make sure you are using the latest revision of the part print and process plan?
5. Explain the difference between part program check-out and part program prove-out.
6. The amount of time spent on a part program prove-out will depend on four things. Name them.
7. A _____ run means that the *program* is loaded into the control and executed but no *part* has been loaded.
8. Single block mode is controlled by a G code. (True/False)
9. List three general safety rules to follow to have a safe prove-out and production run.
10. OD means _____ and ID means _____.
11. On a CNC turning center, a _____ will move the tool toward the chuck, a _____ will move the tool away from the chuck, a _____ will move the tool toward the spindle centerline, and a _____ will move the tool away from the spindle centerline.
12. Describe the following words on a CNC turning center:
    G _____
    X/Z _____
    M _____
    F _____
    S _____
    T _____

13. List one advantage and one disadvantage each for a vertical and a horizontal machining center.
14. Describe some improvements that have come about as a result of machining center innovations and improvements.
15. Answer True/False for the following sequence.
    - G01 is a canned cycle. _____
    - The R word is only used on CNC turning centers. _____
    - M30 is the miscellaneous function for single block mode. _____
    - A +Z move moves the tool into the work. _____
    - CSS speeds up the rpm on a vertical machining center. _____
16. Name the four basic elements necessary to program circular interpolation.
17. Fill in the meaning of the following:

    G02 _____

    G03 _____

    I word _____

    J word _____

    K word _____

    G40 _____

    G41 _____

    G42 _____
18. In your own words, explain what tool offsets are and what their four basic uses are.
19. _____ permits oversize and undersize cutters to be used while still maintaining the programmed part geometry.
20. When programming a part for any milling operation, the operator/programmer must take three main factors into consideration. Name them.
21. Rotary table motion on a horizontal machining center is designated by the _____ which must be aligned the same as the X-, Y-, or Z-axes.
22. What is *probing* on a CNC machine and how is it used?

23. What are the three main parts of the surface-sensing probe?
24. Using full sentences and in your own words, write a paragraph describing the importance and value of having and using cutter diameter compensation.

SHOP ACTIVITIES

Now that you have been exposed to the math and basic programming, it's time to begin doing some CNC programming and operation in your shop. Remember that many machines and controls are different. Each has its own programming manual. You should now obtain a copy of a programming/operation manual for a CNC machine in your shop and try to do some off-line or at-the-machine programming—with your instructor's or supervisor's permission. Perhaps you can team up with someone else in your class to work together and learn together as you get started. However, *never* attempt to program or operate any CNC machine or control without first consulting the manual for that specific machine and control combination. Good luck!

SECTION 5

TOOLING AND FIXTURING FOR CNC MACHINES

OBJECTIVES

After studying this section, you will be able to:

- Discuss the overall importance of tooling and proper application practices to a successful CNC installation.
- Explain how the correct use of cutting tools affects overall machine performance and productivity levels.
- Describe how to correctly use and care for toolholders and carbide inserts.
- Discuss clamping, work holding, and setup tooling for CNC machines.
- Identify some sound tooling practices.

Tooling and fixturing are very important to the success of any CNC installation!

INTRODUCTION

In this section you will learn about the overall importance and proper application of tooling on CNC machines. You may have used and studied cutting tools, fixtures, toolholders, and carbide inserts before. But there are some additional things to think about and be aware of with tooling on CNC machines.

CNC machines merely position and drive the cutting tools by way of the program that you or someone else has written. It's the actual cutting tool that cuts the part to the required print dimensions. Cutting tools take on a heightened level of importance on CNC machines because they are "in the cut" longer. That is, once a program is written and proved out on the machine, the process is basically automated. The operator's job then is to watch over the machine and process, load and unload parts, and replace worn or damaged tooling. Consequently, cutting tools and inserts wear out and are used up faster than on manual machines. Part and fixture location, clamping, toolholders, inserts, and good, safe machining practices are also important to successful CNC operation.

The CNC operator has just as much responsibility for producing quality work on CNC machines as he or she does on manual machines. The operator must not be lulled into a false sense of security because "the machine runs from the program and someone else wrote the program." Paying close attention is required at all times relative to accurate part location, distortion under clamping pressure, watching and listening for abnormal machining conditions, and other tooling related conditions.

As you read and study this section, keep in mind that the best, most expensive, and most accurate CNC machine made is only as good as the tooling and tooling practices used with it.

UNIT 27

Cutting Tools and Application Practices

KEY TERMS

perishable tools
subland drills
spiral pointed taps
length-to-diameter ratio (L/D)

drill tolerances
hand taps
torsional rigidity

machine screw taps
right- or left-hand helix
spiral-fluted taps

CUTTING TOOL CONSIDERATIONS

Tooling for N/C and CNC machines has always been one of the most neglected elements of an N/C installation. When planning, justifying, and even using N/C equipment, all tooling tends to be taken for granted—until something goes wrong.

N/C machines can only move or position cutting tools to specific locations and rotate them or the workpiece at desired spindle speeds. The individual cutting tools actually do the metal removal work. The only way a CNC machine can be efficiently and effectively used is through proper use and care of cutting tools and work-holding devices.

In conventional machining, part accuracies depend on special fixturing. This type of fixturing has precisely made, precisely located tool-setting pads and accurately located bushings that guide the tools. With a CNC machining center, simple fixturing is used. There are no tool bushings or tool pads to guide the tools. The repetitive positioning accuracy of the machine promises a high degree of quality. However, the accuracy of the finished part is only as accurate as the cutting tools and holders used to produce the part. If a drill "runs out," the benefit of the machining center's accuracy is lost. The programmer must assume that the tools will not run out.

There is another reason for careful selection of cutting tools. The average, conventional machine tool cuts metal only 20% of the time. A CNC machining center can be expected to cut metal up to 75%, or more, of the time. This results in more tool usage in a given period of time. Tool life, measured in "time in the cut," will be as good or better but, because of the increased usage, cutting tools will be used up three times as fast. The cost of *perishable tools* (tools that can be used for only a limited period of time and then are thrown away; e.g., some drills, taps, reamers, end mills, etc.) used during the machine's lifetime may amount to 50% or more of the machine's purchase price. Therefore, perishable tools represent a sizable tooling investment over the life of the machine.

A variety of cutting tools are used on N/C equipment to perform a multitude of machining operations. Many of the cutting tool applications, however, are no different than that which would have to be performed on manual equipment to produce the same workpiece. Cutting tools range from conventional drills, taps, and end mills to high-technology carbide cutting tools. Because of the importance of cutting tools to the overall manufacturing process, as well as their cost, it is important that each be examined in detail.

> **"Tech Talk"**
>
> Machinability describes the ease or difficulty with which a metal can be machined. Such factors as cutting tool life, surface finish produced, and power required must be considered.

DRILLS

All new drills have certain allowable *drill tolerances,* as shown in Figure 27–1. The drill tolerances that affect accuracy the most are lip height, web centrality, and flute spacing. The lip height, for example, of a .250-inch drill can vary 0.004 inch, its web can be off center as much as 0.005 inch, its flute spacing can be off by 0.006 inch, and the drill will still be within required specifications. Since a .250-inch drill is normally fed at a rate of .004 or .005 ipr, it would be impossible for that drill to produce accurate holes.

One of the most important things to think about when selecting a drill is to choose the shortest drill length that will permit drilling the hole to the desired depth. A good rule to remember is: *the smaller the drill size, the smaller the allowable error;* as drill size increases, the allowable error progressively increases. Short, stubby drills run truer, allow the fastest feeds, and improve tool life. The torsional rigidity of a drill will affect not only tool life and feed potential, but hole quality as well. *Torsional rigidity* is a measure of the tool's abil-

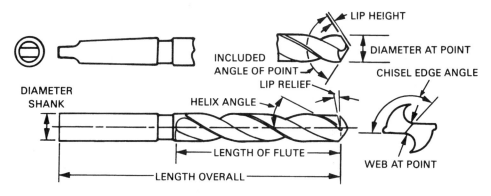

FIGURE 27-1
Identification of toleranced areas for two-flute general purpose drills.

ity to resist twisting or unwinding; rigidity increases as drill length decreases. Therefore, on machining centers, where feed is constant and rapid, a shortened flute is a real advantage.

There are many different types and varieties of drills used for a wide variety of applications. The common twist drill certainly has its applications but so do center drills, spade drills, and subland drills. Center drills, or spotting drills, are primarily designed to produce accurate centers in the work so that follow-up drills will start in perfect alignment. The proper selection and use of these drills will increase the accuracy of hole location, particularly on rough surfaces.

Ideally, the center drilled hole should be machined to a depth where the countersunk portion is 0.003 to 0.006 inches larger than the finished hole size (Figure 27-2). With this method, the drill periphery will be guided into the countersunk hole, the location will be accurate, and the finished hole will have a chamfered or deburred edge.

If large-diameter holes are required to be machined, the twist drill may be impractical. Spade drills, as illustrated in Figure 27-3, are sometimes considered because they can produce large holes in one pass. Conventional twist drills can also be used for larger holes by progressively increasing the drill sizes until the desired size is obtained.

Spade drills have an advantage in N/C work because only the blade, not the entire tool, needs to be changed when it becomes dull.

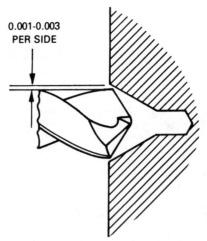

FIGURE 27-2
Ideal center drill size in relation to finish hole size.

Thus, correct tool length is maintained and resetting the tool length or recompensating the machine is eliminated.

The spade drill will normally use the same feeds and speeds as a twist drill. In cast iron, the spade drill performs well at almost any depth. However, in steel and aluminum, if the hole depth is more than one and one-half or two times the hole diameter, problems with

FIGURE 27-3
Spade drill showing blade and shank. (Courtesy of DoAll Company.)

FIGURE 27-4
A multiple-diameter (subland) drill. (Courtesy of Cleveland Twist Drill.)

heat and chip removal can occur. Because of the advances made in carbide insert tooling, spade drills are not used as much as they were years ago.

Many hole-producing sequences require multiple operations on the same hole, such as drill and countersink, drill and counterbore, or drill and body drill. Multiple-diameter drills, called *subland drills*, are commonly used today (Figure 27-4). The proper use of this type of drill can result in a time savings and quality improvement. By combining multiple drilling operations into one tool, extra machining time and tool handling time are eliminated. An additional benefit is derived from the rigidity of the larger diameter, both in the ability to use higher feed rates and in improved hole accuracy.

TAPS

Tapping is one of the most difficult machining operations because of the ever-present problem of chip clearance and adequate lubrication at the cutting edge of the tap. This is further aggravated by coarse threads in small diameters, tough materials, and countless other factors. Further, the relationship between speed and feed is

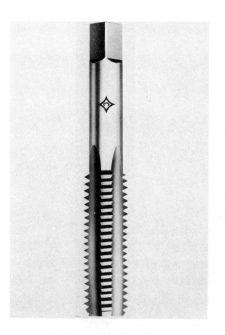 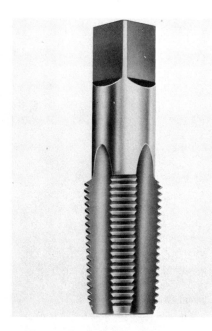

FIGURE 27–5 A B
Conventional hand taps (A and B). ([A], Courtesy of Cleveland Twist Drill. [B], Courtesy of Morse Cutting Tools.)

fixed by the lead of the tap and cannot be changed independently. Tapping on a CNC machine frees the operator of the skill needed to tap a good hole since the tapping operation is programmed. Therefore, the prime concern is the tap, not the skill.

Generally, taps are divided into two major classifications: **hand taps** and **machine screw taps.** Their names, however, do not denote the manner in which the taps are used, because they are both used in power tapping of drilled holes.

Hand taps (Figure 27–5) were originally intended for hand operation but now are widely used in machine production work. The name applies to the group of taps that are available in fractional sizes. The most commonly used hand taps include sizes ranging from 1/4 to 1½ inches.

Machine screw taps are a group of taps available in decimal sizes. Machine screw taps are actually small hand taps. Their size is indicated by the machine screw system of sizes, ranging from #0 to #20.

Spiral pointed taps (Figure 27–6) are sometimes referred to as *gun, chip-driver,* or *cam-point* taps. They are recommended for through

FIGURE 27-6
Spiral pointed, gun, or chip driver tap. (Courtesy of DoAll Company.)

holes and for holes with sufficient clearance at the bottom to provide chip space. Spiral pointed taps have straight flutes with a secondary grind in the flutes along the chamfer. This is ground into the flute at an angle to the axis of the tap so that it produces a shearing action when cutting the thread. As a result of this shearing action, the chips are forced ahead of the tap with very little resistance to thrust.

The main advantage of the spiral pointed tap is that it prevents chips from packing in the flutes or wedging between the flanks and the work. This is a major cause of tap breakage, particularly in small taps. The spiral pointed tap also allows a better flow of lubricant to the cutting edges.

Spiral-fluted taps (Figure 27-7) are recommended for tapping blind holes where the problem of chip elimination is critical. They are most effective when the material being tapped produces long, stringy, curling chips. The spiral-fluted tap cuts freely while ejecting chips from the tapped hole. This prevents clogging and damage to both the threads and the tap. Chip removal is accomplished by the backward thrust action of the spiral flutes.

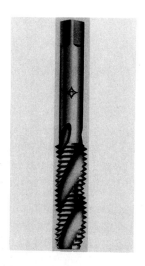

FIGURE 27-7
Spiral fluted or turbo tap (A and B). ([A], Courtesy of Cleveland Twist Drill. [B], Courtesy of DoAll Company.)

FIGURE 27-8
Fluteless taps. (Courtesy of DoAll Company.)

In tapping operations, there are many factors which reduce tap life. Studies of the relating factors indicate that most problems stem from one area: eliminating chips from the hole. However, with fluteless taps (Figure 27–8), this problem and others can be solved. Fluteless taps do not produce chips. Rather, they form or roll the threads into the hole through cammed lobes on the periphery of the tap. Because of this forming action, the tap drill used is always larger than that used for a conventional tap.

Traditionally, certain materials have been tapped dry; plastics and cast iron, for example. Even where dry tapping is possible, improved performance is generally effected by a careful selection of some type of lubricant. Tapping lubricants serve several purposes, the most important being:

- They reduce friction.
- They produce clean, accurate threads by washing chips out of the tap flutes and the threaded hole.
- They improve the thread's surface finish.
- They reduce build-up on edges or chip welding on the cutting portion of the tap.

REAMING

Reaming is the process of removing a small amount of material, usually .062 inch or less, from a previously produced hole. The reamer is a multibladed cutting tool designed to enlarge and finish a hole to an exact size. Since the reamer is basically an end cutting tool mounted on a flexible shank, it cannot correct errors in hole location, hole crookedness, etc. A reamer will follow a previously produced hole. Therefore, when straightness or location is critical, some prior operation, such as boring, must have been performed in order to obtain these qualities.

Reamers can have either straight or spiral flutes and either a *right- or left-hand helix* (a twist of the blade to the left or right) (Figure 27–9). Those with spiral or helical flutes will ordinarily provide smoother shear cutting and a better finish.

The shell reamer (Figure 27–10) is primarily used for sizing and finishing operations on large holes, usually 3/4 inch and larger. The

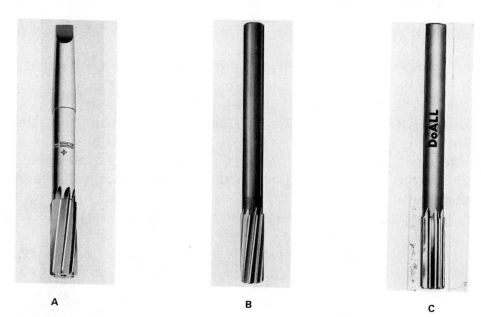

FIGURE 27-9
Some typical reamers: spiral and straight flutes (A, B, and C). ([A], Courtesy of Cleveland Twist Drill. [B], Courtesy of Morse Cutting Tools. [C], Courtesy of DoAll Company.)

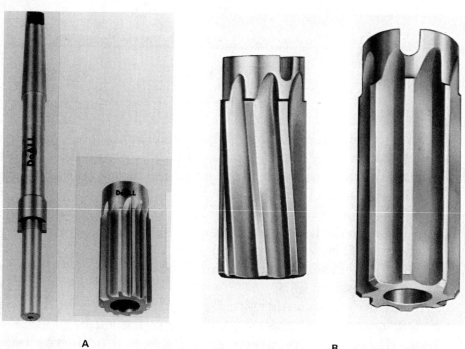

FIGURE 27-10
Shell reamers (A and B). ([A], Courtesy of DoAll Company. [B], Courtesy of Morse Cutting Tools.)

reamer will fit either a straight or taper shank arbor. Several different sizes of shell reamers may be fitted to the same arbor. This results in a tool savings.

Reaming on a CNC machine is an accepted practice. Floating toolholders usually are not required for reaming operations on a CNC machining center for two reasons. First, the repetitive positioning accuracy of the machine slides ensures that the reamer will be positioned in line with the hole. Second, the accuracy of present-day collets and adaptors ensures that the reamer will run "true," and therefore cut to size.

BORING

Boring operations are performed to produce accurate diameters, accurate locations, good finishes, and true, straight holes. Properly performed, boring is the one hole-finishing process whereby the full positioning accuracy of a CNC machine can be used. A cored hole may be cast out of location or, in drilling, the drill may wander beyond the acceptable tolerance. Boring can correct these errors and finish the hole with a high degree of accuracy.

A variety of boring bars, similar to the ones in Figure 27–11, and numerous types of cutters are used for modern boring operations. Regardless of the type used, all have certain common characteristics. Every boring cutter has a side cutting edge and an end cutting edge. These edges are related to the tool shank and are part of the standard nomenclature of single-point cutting tools of the American National Standards Institute (ANSI).

An important element of boring for you to consider is the *length-to-diameter ratio (L/D)* of the boring bar. This ratio is one of the most important, but most neglected and least understood, aspects of boring. It refers to the length of the boring bar in relation to its diameter. Some of the largest manufacturers of boring machines and boring bars have researched this problem extensively. Studies indicate that a boring bar with a 1:1 length-to-diameter ratio is 64 times more rigid than a boring bar with a 4:1 ratio and 343 times more rigid than one with a 7:1 ratio. It follows, then, that in order to obtain the best rigidity and accuracy from boring operations, the boring bar should be as short as possible.

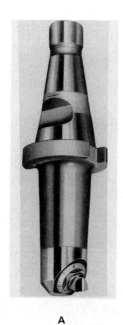

FIGURE 27–11
Typical boring bars (A and B). (Courtesy of Cincinnati Milacron, Inc.)

MILLING

With the exception of drills, probably the most widely used, efficient, and economical metal-removal tool for a machining center is the end mill. Since machining centers are most efficient for short- to medium-sized production runs, it is evident that an end mill, similar to the one in Figure 27–12, takes its place as one of the basic tools for machining centers.

Considering the work potential of end mills, a machining center's contouring capability should be used often. Many parts have milled surfaces, bored holes, recesses, counterbores, face grooves, and pockets; all of these operations are relatively easy to perform with two- or three-axis contouring on most modern machining centers. In many cases, bored holes can be rough- and semifinish-bored by programming an appropriate end mill in a circular path around the centerline of the bore. This method can result in some very good savings. One end mill can replace two or more boring bars, and one end mill can be used for several different bores. Tool drum storage space can be freed for additional tools, tool inventory is reduced, and some time can be saved.

CUTTING TOOLS AND APPLICATION PRACTICES • 267

FIGURE 27-12
Double-end end mill. (Courtesy of Precision Industries, Inc., Providence, RI.)

Many other types of end mills may be used on N/C machines, such as shell end mills with serrated and indexable blades, and face mills for a variety of applications.

COUNTERSINKING AND COUNTERBORING

Countersinking on CNC machines can be a frustrating experience because the pointed angle (Figure 27-13) makes it difficult to establish the tool set length. Consequently, countersink depths may be too shallow or too deep. To set up a job accurately using a standard, single-flute countersink, optical measuring equipment must be used.

Counterboring operations typically are done with a three- to eight-fluted counterbore (Figure 27-14). Counterbores are designed with fixed or removable pilots to produce counterbores concentric with the drilled hole. With the repetitive positioning accuracy of most

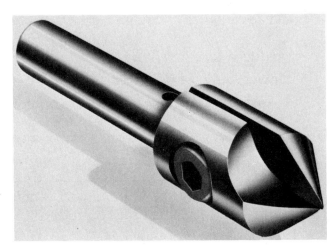

FIGURE 27-13
Standard, single-flute countersink. (Courtesy of Precision Industries, Inc., Providence, RI.)

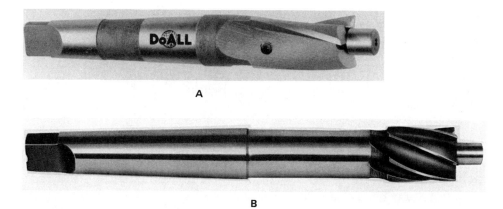

FIGURE 27-14
Typical piloted counterbore. ([A], Courtesy of DoAll Company. [B], Courtesy of Cincinnati Milacron, Inc.)

CNC machines, however, the need for piloted or hole-guided counterbores is often eliminated. Standard counterbores may be removed from the tooling for machining centers, except in long-reach applications.

Most counterboring in shops with machining centers is done with end mills. It is good machining practice to make the necessary end mills from counterbores because end mills ranging from 3/16 inch to 2 inches are readily available. Doing so will yield a large variety of counterbores with almost infinite size availability, and the faster helix means greater feed rates and faster chip removal.

Some advantages of using end mills as counterbores are:

- reduced tool inventory
- lower tool cost per piece
- ease in producing a spotface or counterbore on a rough or angled surface.

RECALLING THE FACTS

1. The _____ of the finished part is only as _____ the cutting tools and holders used to produce the part.
2. _____ are tools such as drills, taps, reamers, etc. that have a limited life and are eventually thrown away.

3. The _____ that affect accuracy the most are lip: height, web centrality, and flute spacing.
4. Multidiameter drills are called _____ drills.
5. _____ taps apply to the group of taps that are available in fractional sizes; _____ taps is the name given to the group of taps available in decimal sizes.
6. _____ taps are sometimes referred to as _____, *chip-driver*, or *cam-point* taps. They are recommended in N/C work for _____ holes and for holes where enough chip space exists at the bottom of the hole.
7. The difference between spiral pointed taps and spiral-fluted taps is _____
   _____
   _____ .
8. Tapping lubricants help reduce _____, produce clean, accurate threads by washing _____ out of the tap _____ and threaded hole, improve the _____ surface finish, and reduce thread _____ on the cutting portion of the tap.
9. A _____ is a twist of the cutting blade to the left or right.
10. The _____ ratio in boring refers to the length of the boring bar in relation to its diameter.
11. With the exception of drills, the most widely-used, efficient, and economical metal-removal tool for a machining center is the _____ .
12. Advantages of using _____ _____ as counterbores are that they reduce tool _____, lower the tool _____ per _____, and more easily produce a _____ or _____ on a rough surface.

UNIT 28

Using Carbide Inserts and Toolholders

KEY TERMS

IC (inscribed circle)
coated carbide inserts
titanium-nitrite (TIN)

combination drills
ceramic inserts
chipping

coated inserts
cratering

CARBIDE INSERTS

Carbide inserts are widely used in manufacturing for both manual and CNC machining. They are available in many standard sizes and shapes affecting both strength and cost. The relationship of strength and cost for standard sizes and shapes of carbide inserts can be seen in Figure 28–1. Insert size is measured by the largest circle that will fit entirely on the insert called the *IC (inscribed circle)*. Standard inserts have an IC of 1/4, 3/8, 1/2, or 3/4 inch.

Insert shape is commonly marked or indicated by a letter, such as "T" for triangular, "R" for round, etc. Inserts also have different nose radii for different applications. Generally, the smaller the IC of an insert, the less it costs. The nose radius of an insert affects the surface finish and the maximum allowable feed rate. Large-radius inserts usually leave a better surface finish than small-radius inserts. Also, large-radius inserts have a stronger cutting edge and throw off heat better, allowing higher feed rates. Inserts are also identified by the industry standard identification system chart shown in Figure 28–2. As you study this chart, you will see that each position in the insert identification number specifies a position of the insert's geometry.

Carbide inserts are available in a range of grades, each designed for specific machining applications. Hard grades of carbide will contain a higher percentage of tungsten and are less shock-resistant. Heavy roughing cuts require a softer shock-resistant grade of carbide, while finishing cuts require a harder wear-resistant grade. Additionally, the material being machined also determines the

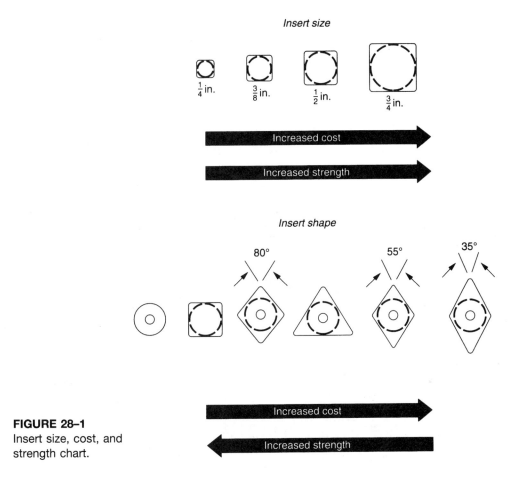

FIGURE 28-1
Insert size, cost, and strength chart.

choice of insert grade. Nickel-based alloys are very tough and need a soft grade of carbide while abrasive cast irons usually call for a very hard grade of carbide.

Other types of materials are used as the base material for inserts and for coating carbide inserts. *Ceramic* (cemented oxide) inserts are used today for a variety of materials and in a variety of machining applications. *Coated carbide inserts,* with a *Titanium-nitrite (TIN)* coating, are also widely used today. The extremely hard and wear-resistant TIN coating is bonded to a soft, tough carbide base, giving the coated insert the best properties of both hard and soft insert grades. TIN coating, usually yellow or gold in color, allows a single insert to be used over a wider range of applications.

Some carbide insert tools such as *combination drills,* similar to the one shown in Figure 28–3, can be used for facing and turning

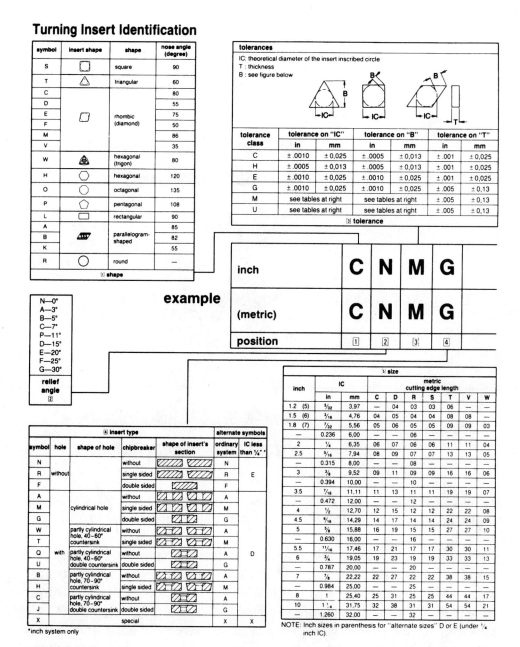

FIGURE 28-2
Industry standard carbide identification system. (Courtesy of Kennametal Inc.)

USING CARBIDE INSERTS AND TOOLHOLDERS • 273

IC		† tolerance on "IC"							IC		± tolerance on "B"						
		class "M" tolerance				class "U" tolerance					class "M" tolerance				class "U" tolerance		
		shapes S, T, C, R & W		shape "D"		shapes "S, T & C"					shapes S, T, C, R & W		shape "D"		shape "V"		shapes "S, T & C"
inch	metric	inch	mm	inch	mm	inch	mm		inch	metric	inch	mm	inch	mm	inch	mm	inch	mm
5/32	3,97			—	—	—	—		5/32	3,97			—	—	—	—	—	—
3/16	4,76			—	—	—	—		3/16	4,76			—	—	—	—		
7/32	5,56	.002	0,05						7/32	5,56	.003	0,06			—	—		
1/4	6,35			.002	0,05	.002	0,05	.003	0,06	1/4	6,35			.004	0,11	—	—	.005	0,13
5/16	7,94								5/16	7,94					—	—		
3/8	9,52								3/8	9,52					.007	0,18		
7/16	11,11								7/16	11,11					—	—		
1/2	12,70	.003	0,06	.003	0,06	.003	0,06	.005	0,13	1/2	12,70	.005	0,13	.006	0,15	.010	0,25	.008	0,20
9/16	14,29								9/16	14,29					—	—		
5/8	15,88								5/8	15,88					—	—		
11/16	17,46	.004	0,10	.004	0,10	.004	0,10	.007	0,18	11/16	17,46	.006	0,15	.007	0,18	—	—	.011	0,27
3/4	19,05								3/4	19,05					—	—		
7/8	22,22	.005	0,13						7/8	22,22					—	—		
1	25,40			—	—	—	—	.010	0,25	1	25,40	.007	0,18	—	—	—	—	.015	0,38
1 1/4	22,22	.006	0,15	—	—	—	—			1 1/4	22,22	.008	0,20	—	—	—	—		

—	4	3	2	☐	☐	☐
	12	04	08	☐	☐	☐
		5			6			7			8			9			10	

hand of insert (optional) |8|

R   L

|9 & 10|
cutting edge condition or
chip control features (optional)

T — negative land
K — light feed chip control, double sided Kenloc insert
M — heavy feed chip control, deep floor Kenloc
N — narrow land Kentrol insert with chip control on one side
W — heavy-duty chip control, wide land Kenloc insert one side
J — polished to 4-microinch AA (rake face only)
UF — ultra-fine finishing
See Technical Section for additional conditions and chip control features.

|5| thickness

thickness		symbol	
in	mm	inch	metric
1/32	0,79	0.5 (1)	—
1/16	1,59	1 (2)	01
5/64	1,98	1.2	T1
3/32	2,38	1.5 (3)	02
1/8	3,18	2	03
5/32	3,97	2.5	T3
3/16	4,76	3	04
7/32	5,56	3.5	05
1/4	6,35	4	06
5/16	7,94	5	07
3/8	9,52	6	09
7/16	11,11	7	11
1/2	12,70	8	12

NOTE: Inch sizes in parenthesis for "alternate sizes" D or E (under 1/4 inch IC).

|7| corner radius

corner radius		symbol	
in	mm	inch	metric
.004	0,1	0	01
.008	0,2	0.5	02
1/64	0,4	1	04
1/32	0,8	2	08
3/64	1,2	3	12
1/16	1,6	4	16
5/64	2,0	5	20
3/32	2,4	6	24
7/64	2,8	7	28
1/8	3,2	8	32
round insert (inch)		—	00
round insert (metric)		—	M0

FIGURE 28–2
(continued)

FIGURE 28–3
Carbide insert drill-face-turn tool. (Courtesy of Valenite, Inc.)

operations as well as for drilling a hole from solid on a CNC lathe—all with the same tool! In addition, cutting speeds and feeds are greatly increased, thereby improving productivity levels. The high-technology indexable insert (meaning that the insert can be indexed

FIGURE 28–4
High-technology carbide insert drill. (Courtesy of Valenite, Inc.)

> **"Tech Talk"**
>
> Carbide inserts are produced by mixing very pure, finely ground grains of tungsten carbide with a cobalt binder and other additives. This material is packed tightly into molds shaped like the familiar square, round, diamond, or triangle insert shapes, then squeezed under high pressure. The "green" insert is then removed from the mold and heated at high temperature to fuse all of the tightly packed grains into a solid mass. The insert is then ground, cleaned, and may be coated with titanium nitrite (TIN) or other materials.

to several different cutting edges on the same insert) drilling bars (Figure 28–4) were designed to replace conventional twist drills and spade drills. These drills are capable of running up to ten times faster, depending upon material type, because they run at coated carbide speeds and feeds. The indexable insert drilling bar uses multiple-edge, two-sided indexable inserts in most cases. Use of these indexable insert drills reduces machine idle time because only the inserts need to be replaced, not the entire drill. The inserts can be indexed while still at the machine, and the use of inserts eliminates the need for tool resharpening.

Most carbide insert drills have other features which make them extremely advantageous when compared to conventional twist drills. These include:

- Increased web thickness. This gives them the strength to handle high penetration rates.
- Added rigidity and chatter avoidance due to the larger shank diameters.
- Semifinish boring operations can be eliminated. This is due to the fact that a hole can be drilled very close to the final size desired.

Good machining practices are very important with indexable insert tooling. Flying chips, for example, create a danger to the operator and a safety shield is usually required. Coolant is also necessary to cool the cutting edge and to flush away chips. Sufficient coolant must be applied continuously on these tools. This is due to the higher chip removal rates, speeds, and feeds. In addition, drilling with carbide insert drills develops high thrusts. If the setup is not rigid, the forces will create chatter and side loading. This can result in broken inserts and damage to the drilling bar.

FIGURE 28–5
A turning operation showing heat as the primary destroyer of insert life. (Courtesy of Cincinnati Milacron, Inc.)

> **"Tech Talk"**
>
> Tungsten carbide, the chemical compound of tungsten and carbon, is composed of 94 percent tungsten and 6 percent carbon. Tungsten carbide was first discovered by Moissan, a Frenchman, in the late 1800s. Moissan found that by mixing tungsten powder with carbon and heating them to a high temperature, he formed a very hard wear-resistant material.

Heat is the primary cause of insert wear and destroyer of insert life. Shearing action of the metal being formed into a chip (Figure 28–5) and the chip rubbing against the cutting tool are the sources of the heat. Other sources of insert wear include too high or too low feeds and speeds, too much of a depth of cut, and various insert problems caused by tool pressure and stress.

Two additional causes of insert wear are *cratering* and *chipping*. Cratering is generally caused by improper insert selection based on the alloy content of tungsten carbide with titanium and tantalum carbides. The improper alloy content causes a lack of resistance to crater and flank wear resulting in breakdown of the cutting edge. Inserts designed for high-speed applications should contain large amounts of titanium to resist cratering.

Chipping is caused by impact on the insert cutting edge, or vibration and lack of rigidity in the machining setup. Using an insert cutting edge that is not crack- or vibration-resistant, or too hard a grade of carbide for a particular application, promotes chipping. Material may weld itself to the insert cutting edge if cutting speeds are too low. When the built-up material is removed, a portion of the cutting edge may also be removed, thereby causing a small chip in the insert.

Many CNC machining applications require the use of cutting fluids. Coolant usage is an important factor related to cutting tool life because it:

- cools and carries heat from the cutting edge of the tool and workpiece. It is advisable to use a continuous flow of coolant over the cutting edge and entire operation (Figure 28–6), or

FIGURE 28–6
A continuous flow of coolant over the cutting edge of the tool can be important to extending the life of the cutting tool. (Courtesy of Cincinnati Milacron, Inc.)

FIGURE 28–7
A heavy turning center stock-removal application showing chip formation and flow. (Courtesy of Cincinnati Milacron, Inc.)

not use it at all. Rapid heating and cooling of the insert by turning coolant on and off may cause insert heat cracks.
- lubricates and reduces heat friction at the face of the tool and allows smooth chip flow off the cutting edge. This reduces tool wear and power consumption, while improving surface finish.

Cutting fluids have a secondary purpose of "damping down" flying dust or fine particles that are a safety hazard to the health of the operator.

Chip control, often regarded as a metalworking nuisance, is generally handled in a hit-or-miss fashion. However, chip formation and flow (Figure 28–7) are generally functions of proper or improper tooling, feeds, speeds, and depth of cut.

TOOLHOLDERS

Any discussion of cutting tool considerations should include toolholders. Spindle-type toolholders (Figure 28–8) essentially hold and drive the entire assembly, which is made up of the basic cutting tool,

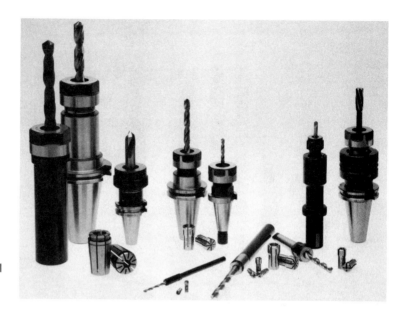

FIGURE 28-8
Spindle-type tool holders hold and drive the entire tool assembly. (Courtesy of Kennametal Inc.)

adapters or collets, and the toolholder itself. Most toolholders conform to the ANSI industry standard for tapered V-flange toolholders.

The most important thing about toolholders is that they be free from nicks, gouges, dirt, grit, and any other visible signs of damage. Lack of toolholder cleanliness can cause the entire tool assembly to "run out," resulting in part inaccuracies. Toolholders must be cleaned, inspected, reconditioned, if possible, or scrapped if accuracy or safety becomes questionable. Toolholders must accurately drive the cutting tools and run true. Also, machine tool spindles and tool matrix pockets should be periodically inspected and wiped out (*while the machine is stopped and in the manual mode*) to remove dirt, chips, and grit. Gouging of the machine spindle or the toolholder can result when pull-back clamping pressure is applied on the toolholder. Accepting anything short of "the best" tool assembly is settling for second-rate performance of the CNC machine tool.

Turning center toolholders using disposable inserts, as seen in Figure 28–9, permit indexing a carbide insert that has six to eight cutting edges in less than one or two minutes. If part tolerances are relatively open, the original tool setting can be maintained and the machine is ready for production with no additional tool adjustment.

FIGURE 28-9
Turning center tool holders that use disposable inserts permit indexing of carbide inserts. (Courtesy of Kennametal Inc.)

RECALLING THE FACTS

1. Insert size is measured by the largest circle that will fit on the insert, called the _____.
2. Generally, the smaller the _____ of an insert, the less it costs.
3. Cemented oxide inserts, called _____, are used for a variety of machining applications today.
4. Inserts today are _____ with a variety of new materials which make the insert more wear-resistant.
5. Some carbide inserts today are bonded with a material known as _____ which makes the insert extremely hard and wear-resistant and suitable for machining multiple types of material.
6. Some carbide insert tools called _____ can be used for facing, turning, and drilling a hole from solid on a CNC lathe.
7. The primary cause of insert wear and destroyer of insert life is _____.

8. _____ is generally caused by improper insert selection based on the alloy content of the carbide material.

9. _____ is caused by impact to the insert cutting edge or vibration and lack of rigidity in the machining setup.

10. Coolant usage is an important factor to cutting-tool life because it _____ from the cutting edge of the tool and the workpiece, and _____ heat friction at the face of the tool and allows smooth chip flow off the cutting edge.

11. Rapid heating and cooling of the insert by turning coolant on and off may cause insert heat _____.

12. Most toolholders conform to the _____ industry for tapered _____ toolholders.

13. The most important thing to pay atention to about toolholders is that they be free from _____, _____, _____, and _____ and any visible signs of damage.

UNIT 29

Fixture Usage and Basic Principles

KEY TERMS

jig *fixture* *hard fixtures*
modular fixturing *foolproofing*

FIXTURE DESIGN AND USAGE

Jigs and fixtures are production tools that help to manufacture duplicate parts accurately. The purpose of jigs and fixtures is to hold, support, and locate every part to ensure that each is identically machined within the specified limits.

Sometimes the term *jig* is used to mean *fixture* and the term *fixture* is used to mean *jig*. It is important for you to know the difference. A *jig* is a part-holding and -locating device that also positions and guides the cutting tool. A *fixture* only holds and locates the workpiece.

Because jigs hold and locate the workpiece as well as position and guide the cutting tool, they are used primarily on manual machines such as drill presses. Generally, small jigs are not fastened to the machine table, but if holes of .25 or larger inch are to be drilled, the jig should be clamped to the table. Jigs are not used on CNC machines. Two examples of jigs are illustrated in Figures 29–1 and 29–2.

Fixtures are used on both manual and CNC machines. Precision set blocks and feeler gauges are used with fixtures on manual machines to correctly position the cutter to the workpiece. With CNC machines, fixtures must accurately hold and locate the part as well as orient it to the machine table. Examples of some different types of fixtures are seen in Figures 29–3 and 29–4.

Proper fixturing is very important to successful CNC machining. A poorly designed or manufactured fixture often causes problems at the machine, holds up production, or produces much waste. As you have already learned, the basic function of a fixture is to locate and secure the part for machining. This involves loading, clamping,

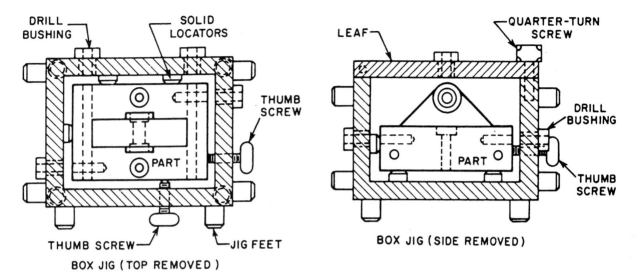

FIGURE 29-1
A typical part-holding jig. (Edward G. Hoffman, *Jig and Fixture Design*, ed. 3 © 1991, Delmar Publishers.

> **"Tech Talk"**
>
> The complexity of a workpiece and processing are the key factors in the cost of making any product. Therefore, a simple change to the design of a workpiece can reduce the manufacturing cost involved.

unclamping, and unloading each workpiece for machining. Loading, clamping, unclamping, and unloading make up an important part of the nonproductive cycle time of each part. If this operation can be simplified, more parts can be produced each hour. Therefore, fixtures should be designed to reduce the overall part setup time. Also, because smaller batch lots run on CNC machines, fixtures are set up on the machine, taken off, and used more frequently. Therefore, tool designers try to design fixtures that simplify the process of loading/unloading, locating, and securing the fixture to the machine table.

The accuracy of all parts that require special fixtures depends on how well and how accurate the fixture is designed, manufactured, and assembled. The advantages of sound, economical design and accurate manufacture and assembly of fixtures are:

- reduced fixture-to-machine and part-to-fixture setup time
- consistency of part accuracies

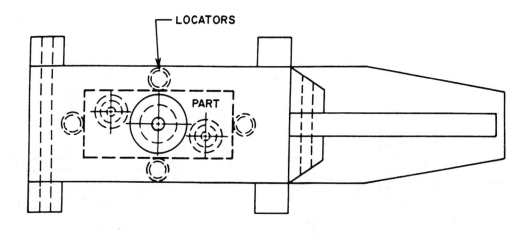
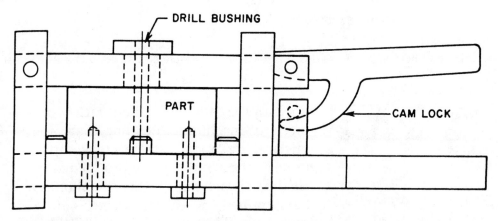

FIGURE 29-2
Jigs hold and locate the part in addition to positioning and guiding the cutting tool. (Edward G. Hoffman, *Jig and Fixture Design*, 3rd ed. © 1991 Delmar Publishers.)

- reduced errors and inaccuracies in part location
- decreased cost per part
- reduced inspection time
- minimal fixture modifications or rework
- faster and easier CNC program prove-out

Another important aspect of fixture design is that of clamping the workpiece. Clamps should always be placed as close to the support locations as possible. Placing unsupported clamps at any convenient location on the fixture could cause the part to cock or distort under

FIGURE 29–3
A typical N/C machining center fixture. (Courtesy of Cincinnati Milacron, Inc.)

FIGURE 29–4
Fixtures only hold and locate the part. (Courtesy of Giddings and Lewis, Inc.)

clamping pressure. Clamp positions, for ease of loading and unloading, should also be given a priority. Clamps must avoid blocking hole locations and milling cuts that may interfere with machining. Tool holders should be considered, particularly when the cutter is engaged in the workpiece, so as not to interfere with axis movements.

Fixtures should always be designed with safety in mind. The operator should always be able to easily reach all clamps and any adjusting screws that may need attention. All sharp corners and projections should be minimized, and chip removal should be easy. The perfectly designed fixture is also the safest for those who must use it.

FIXTURING CONSIDERATIONS

One overlooked aspect of fixturing which sometimes doesn't show up until after the fixture is in production is part orientation. Fixtures should always be designed to prevent incorrect loading and part orientation. This is called *foolproofing* the fixture. Foolproofing a fixture design to prevent incorrect part loading means that there should be only one way the part can be loaded and clamped in the fixture; otherwise the part won't fit in the fixture. Foolproofing is very important to fixture design because carelessness can occur when loading and unloading work is so repetitive. Designing a fixture to be foolproofed takes longer but helps prevent scrapped parts and broken tools.

Other points that must be considered regarding the positioning of the part and fixture in relation to the machine table are:

- *Will all the tools reach?* The maximum and minimum tool lengths should be considered to provide the best placement of the fixture to table and part to fixture.
- *Is the fixture in a foolproof position to the machine table?* Much time and effort is often spent designing a foolproof part-to-fixture relationship, and then the fixture is loaded and clamped on a CNC machining center's table 90 or 180 degrees out of position. This can also cause serious damage to the machine, part, or tooling. Fixtures should always be marked in order to be correctly oriented to the machine table.

- *Is the fixture designed and placed on the machine table to support efficient machining?* Sometimes inefficient CNC programming moves are needed to avoid poorly designed and placed fixture locating pads and clamp assemblies. If the machine has a rotary index table, the part-to-fixture and fixture-to-table relationships should be positioned over the center of rotation. This should provide an equal distance from the center of rotation to each face to be machined. It should also make the programming easier in X-, Y-, and Z-axes.

Modular fixturing makes use of reusable components that can be assembled and torn down based on changing part requirements. Components, such as those seen in Figure 29–5, are manufactured to close tolerances and generally can be assembled on a grid plate based on a simple sketch. Modular fixturing, as seen in Figures 29–6 and 29–7, has the following advantages:

- Fixtures can be assembled and torn down, and components reused.
- Designing each new dedicated-to-one-part fixture, manufacturing the components, inspecting the parts, and rework costs can be substantially reduced or eliminated.
- Elimination of fixture design and component manufacturing saves valuable lead time and reduces coordination problems relative to fixture design, manufacturing, and assembly.
- *Hard fixtures* (those dedicated to only one part or part family) are scrapped when part life becomes inactive; modular components can be salvaged and reused.

FIGURE 29–5
Assorted modular fixturing components. (Courtesy of Qu-Co Modular Fixturing.)

FIGURE 29-6
An example of modular fixturing. (Courtesy of Cincinnati Milacron, Inc.)

FIGURE 29-7
A modular fixture showing component buildup and part location. (Courtesy of Cincinnati Milacron, Inc.)

- If an engineering change is required, hard fixturing requires lead time, rework, and inspection; modular components can be easily and quickly moved or replaced with other modular components to suit the change.
- Hard or dedicated fixturing requires large storage facilities; modular components can be stored in bins and assembled as needed.
- The fixture or tool designer is never locked into his or her design because different fixturing solutions can be tried easily by relocating clamps, stops, and locators.
- Stopping the machine for on-machine table setups can be reduced or eliminated; modular fixtures can be accurately re-created and assembled off-line in a fixture setup area in advance of machining operations and then directly transported, loaded, and clamped to the machine table.

Modular fixtures can be designed by trial-and-error without a formal design on paper. A major advantage of modular fixturing is that the fixtures can be constructed at the last minute, once the part to be machined is available. Many times, better fixtures are built by trial-and-error since the problems associated with how the part sits in the fixture can be seen easily.

RECALLING THE FACTS

1. The purpose of _____ and _____ are to _____, _____, and _____ every part to ensure that each is identically machined within the specified limits.
2. A _____ is a part-holding and -locating device that also positions and guides the cutting tool; a _____ only holds and locates the part.
3. _____, _____, clamping, and unclamping make up an important part of the _____ cycle time of each part.
4. The advantages of sound, economical design and accurate manufacture and assembly of fixtures are:
    - reduced fixture-to-machine and part-to-fixture setup time

- _____
- reduced errors and inaccuracies in part location
- _____
- faster and easier N/C program prove-out
- _____
- _____

5. Clamps should always be placed as close to the _____ _____ as possible.
6. Fixtures should always be designed with _____ in mind.
7. _____ a fixture design to correct incorrect part loading means that there should be only one way a part can be loaded and clamped in the fixture.
8. _____ makes use of reuseable components that can be assembled and torn down based on changing part requirements.
9. _____ fixtures are those dedicated to only one part or part family.
10. Advantages of _____ fixturing are:
    - fixtures can be torn down and components reused.
    - _____.
    - _____.
    - _____.
    - _____.
    - _____ or _____ fixturing requires large storage facilities; modular components can be stored in bins and reused.
    - _____.

UNIT 30

Clamping, Workholding, and Setup Tooling

KEY TERMS

safety
spring-loaded L-clamps
heel blocks
rigidity

support
universal clamps
jacks/blocks

plain bar clamps
U-clamps
T-bolts

FUNDAMENTALS OF SETUPS

Safety and *rigidity* are the most important items to consider when locating and securing parts for machining operations. Safety is of primary importance because carelessness in setting up and clamping parts can cause operator injury, broken tools, and scrapped parts. It is the operator's responsibility to control the process and produce a quality workpiece. Unsafe setups, clamping, and part location problems should be reported immediately to shop supervision and programming.

Great forces are exerted in machining operations. Depending on feed rates, spindle speeds, and how a particular part is being held and machined, setup rigidity should be checked before machining begins. An unsecured or nonrigid setup can cause holding fixture, cutting tool, and possible machine damage in addition to a scrapped part. A well-secured and rigid setup reflects the operator's mechanical abilities and good judgement.

Good setup practices begin with the use and care of proper clamping and workholding surfaces. Some of the more common types of clamps used in both manual and CNC machining are:

1. *Plain bar clamps* (Figure 30–1). This simple type of clamp is not suitable when the rapid loading of a workpiece is required.

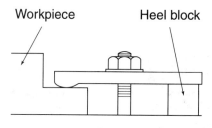

FIGURE 30-1
Simple plain bar clamp.
(From Luggen, William: *Introduction to N/C and CNC* © 1986, Prentice Hall, Upper Saddle River, NJ, pp. 199–203, reprinted with permission.)

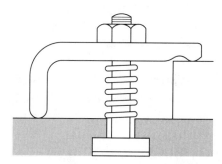

FIGURE 30-2
A spring-loaded L-clamp.
(From Luggen, William: *Introduction to N/C and CNC* © 1986, Prentice Hall, Upper Saddle River, NJ, pp. 199–203, reprinted with permission.)

2. *Spring-loaded L-clamps* (Figure 30–2). Useful when a part can slide easily under the clamps. The spring supports the clamp when the workpiece is removed.

3. *U-clamps* (Figure 30–3). These clamps can be adjusted easily to change the span from workpiece to clamping bolt. The most useful feature is that they can be easily withdrawn or slid aside to make part removal easy. The offset type of U-clamp allows the securing bolt and nut to be below the upper level of the clamp, thus reducing the amount of packing that may be required.

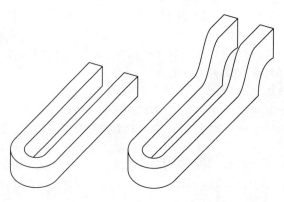

FIGURE 30-3
A U-clamp. (From Luggen, William: *Introduction to N/C and CNC* © 1986, Prentice Hall, Upper Saddle River, NJ, pp. 199–203, reprinted with permission.)

FIGURE 30–4
A universal type clamp. (From Luggen, William: *Introduction to N/C and CNC* © 1986, Prentice Hall, Upper Saddle River, NJ, pp. 199–203, reprinted with permission.)

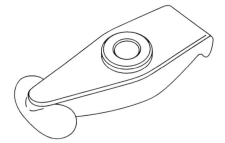

FIGURE 30–5
Height adjustment packers or heel blocks. (From Luggen, William: *Introduction to N/C and CNC* © Prentice Hall, Upper Saddle River, NJ, pp. 199–203, reprinted with permission.)

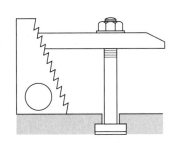 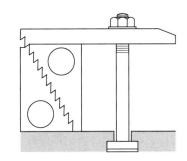

4. *Universal clamps* (Figure 30–4). These clamps allow for height and two-way angle adjustment.
5. *Heel blocks* (height adjustment packers) (Figure 30–5). These can be used singly or in pairs. When used singly, the clamp must have a machined end that suits the "steps." When used in pairs, faces can be interlocked, thus providing various packing heights.

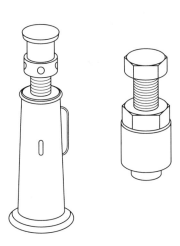

FIGURE 30–6
Support jacks and blocks. (From Luggen, William: *Introduction to N/C and CNC* © 1986, Prentice Hall, Upper Saddle River, NJ, pp. 199–203, reprinted with permission.)

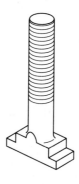

FIGURE 30–7
T-bolts locate in the machine table T-slots. (From Luggen, William: *Introduction to N/C and CNC* © 1986, Prentice Hall, Upper Saddle River, NJ, pp. 199–203, reprinted with permission.)

6. *Support jacks and blocks* (Figure 30–6). These items are used to prevent workpiece deflection under machining forces.
7. *T-bolts* (Figure 30–7). These bolts, which locate in the worktable T-slots, are machined so that they do not turn while the nut is being tightened.

CLAMPING PROCEDURES

Machine tables, vises, and fixture-mounting surfaces should be clean and free from dirt, chips, and burrs. Examine the workpiece and locating surfaces for flattened chips (Figure 30–8) and other foreign matter to ensure positive and square location and mounting. In addition, locating keys

> **"Tech Talk"**
> Grinding, according to some, is the dirtiest of manufacturing processes. Although the process itself is relatively clean, the environmental laws regulating disposal of the sludge waste in landfills are very rigid. Grinding waste in landfills can potentially contaminate the soil and groundwater for decades. Work continues toward finding environmentally-friendly solutions to disposing the waste of this critical manufacturing process.

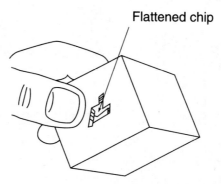

FIGURE 30–8
Part and fixture locating surfaces should be examined for flattened chips, dirt, burrs, and other foreign matter. (From Luggen, William: *Introduction to N/C and CNC* © 1986, Prentice Hall, Upper Saddle River, NJ, pp. 199–203, reprinted with permission.)

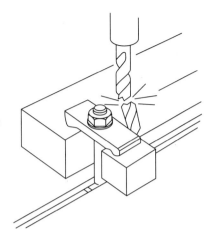

FIGURE 30–9
Collision can result if specific directions for clamping are not followed. (From Luggen, William: *Introduction to N/C and CNC* © 1986, Prentice Hall, Upper Saddle River, NJ, pp. 199–203, reprinted with permission.)

for vises, fixtures, and setup blocks should be checked prior to each use for dirt, chips, and burrs and replaced if required. Exercise care not to damage any locating device when positioning the workpiece against it. Use T-clamping bolts of a length that will permit the nut to be fully engaged on the bolt thread but not stick up more than 1.00 inch above the nut. Make sure that all workholding devices are positioned as specified by the programmer. Use of the wrong equipment or clamping arrangement may cause a collision as the tool positions itself around the workpiece (Figure 30–9).

Some general clamping guidelines are:

- Position the clamping bolts as close to the workpiece as possible (Figure 30–10). Make sure that the heel of the clamp is

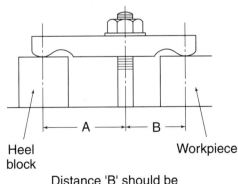

FIGURE 30–10
Clamping bolts should always be positioned as close to the part as possible. (From Luggen, William: *Introduction to N/C and CNC* © 1986, Prentice Hall, Upper Saddle River, NJ, pp. 199–203, reprinted with permission.)

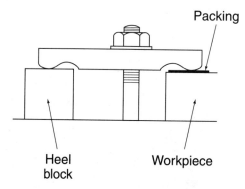

FIGURE 30–11
Soft metal packing should be used between the clamp and the part to prevent damage to the part. (From Luggen, William: *Introduction to N/C and CNC* © 1986, Prentice Hall, Upper Saddle River, NJ, pp. 199–203, reprinted with permission.)

always level or slightly above the the top surface of the workpiece.

- To prevent damage to the workpiece from clamping pressure, insert a piece of soft metal or other suitable material between the clamp and the workpiece (Figure 30–11), but make sure the packing does not extend into the machining path of the tool.

- Tighten clamp nuts, in turn and in equal increments, until the workpiece is secure (Figure 30–12). Before beginning the machining cycle, check that there is no possibility of the workholding equipment interfering with the cutting tool or head of the machine. Failure to pay attention to clamping guidelines and failure to position clamps properly or in the specified location can cause machine, part, or tool damage as well as operator injury.

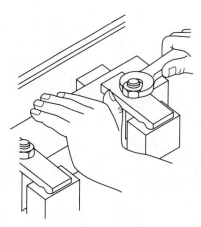

FIGURE 30–12
A part can be distorted if clamp nuts are not tightened in sequence and to the same tightness. (From Luggen, William: *Introduction to N/C and CNC* © 1986, Prentice Hall, Upper Saddle River, NJ, pp. 199–203, reprinted with permission.)

SETUP TOOLING

Location of a particular setup or placement of a fixture on the machine table is another important consideration that must not be overlooked. Table setups, vises, and fixtures should be placed in an appropriate position on the machine table so that all cutting tools are able to reach any part of the workpiece. Following this simple rule will help to reduce bad machining practices resulting from extra-long tools that may be necessitated by poor placement of the setup.

Machine operators should thoroughly inspect a new setup prior to beginning a machining operation. All nuts and bolts should be tightened and all clamps, pushers, and locators should be in place and securely holding the workpiece against positive stops. Setups should be inspected at the beginning of each shift change. For example, many accidents occur because a second shift operator assumed the first shift operator had tightened all the clamps prior to shift change.

RECALLING THE FACTS

1. _____ and _____ are the two most important items to consider when locating and securing parts for machining operations.
2. Setup _____ should always be checked before machining begins.
3. Some of the more common types of clamps used in manual and CNC machining are:
    a. _____
    b. _____
    c. _____
    d. _____
    e. _____
    f. _____
    g. _____
4. Clamping bolts should always be positioned as _____ to the workpiece as possible.
5. Setups should always be inspected at the _____ of each shift change.

UNIT 31

Some Sound Tooling Practices

KEY TERMS

cutting speed **Brinell hardness** *chip per tooth*

Regardless of whether a manual, N/C, or CNC machine is to be used to cut metal, it is very important for you to follow sound tooling practices. Remember, *you* are responsible for the safety and protection of yourself and your co-workers, the accuracy and quality of the work you are producing, and the protection of the machines, tools, and fixtures you are working with.

Cutting tools, as you have learned, play a very important part in the success of any CNC installation. Material characteristics, the type and rigidity of the setup used, how the part is programmed, and a host of other factors can all affect the successful use of those cutting tools. Each programming and machining situation you are faced with is different and requires a careful study and knowledge of all the machining related factors. The following are some general CNC tooling practices, reminders, and guidelines that you need to be familiar with.

> **"Tech Talk"**
>
> When cutting tools cut metal, temperatures at the area of contact between the chip and the tool can run from 1,500 to 2,000 degrees Fahrenheit at the hottest point.

1. *Check all tools* before they are used. This includes the cutting edges and tool body as well as holders, extensions, adapters, etc. They should all be in perfect order and able to act as a total tool assembly.
2. *Select the right tool* for the job and use the tool correctly. Tools should be able to machine the workpiece to the desired accuracy. In purchasing cutting tools and toolholders, examine cost per piece for the part produced as well as the cost of the

tooling package. In many cases, bargain tools cost more per part produced (because you have to use more of them).

3. *Always choose the shortest drill length* that will permit drilling the hole to the desired depth. The smaller the drill size, the smaller the allowable error; as drill size increases, the allowable error progressively increases.

4. *Check and maintain correct **cutting speeds** (the distance in either surface feet or surface meters that a point on the circumference of a tool travels in one minute) and feeds for all tools.* This idea can spell success or failure for any CNC installation. The desired feeds and speeds may not always be achievable, depending on machinability, material characteristics, rigidity of setup, etc.

5. *Understand machine and control capabilities.* Additional tools are often purchased because built-in machine/control capabilities, such as contouring, are not used effectively. Most controllers sold are of the contouring type, but qualified personnel must still be able to recognize where and how their capabilities can be used to avoid tooling problems.

6. *Care for tools properly.* This includes perishable tools, holders, drivers, collets, extensions, etc. They represent a sizeable investment and should be adequately stored and reconditioned when required.

7. *Watch and listen for abnormal cutting tool performances.* Attention paid to actual metal-removal processes can often prevent tool breakage problems, scrapped parts, and rework. Chatter and other vibration problems can be corrected if they are discovered early enough.

8. *Use correct tap drills.* Often the wrong drill is used for a certain size tap, and taps are sometimes broken as a result of negligence and incorrect tool selection.

9. *Holes to be tapped should be deep enough, free of chips, and lubricated prior to tapping.* Insufficient or incorrect lubrication can cause tap breakage, oversized threads, and poor surface finish.

10. *Use end mills as counterbores* where possible. They reduce inventory, produce a lower tool cost per piece, and produce a

spotface or counterbore on a rough or angled surface more easily.

11. *The length-to-diameter ratio should not be exceeded on boring bars.* The boring bar length should never exceed four times the bar diameter. Failure to comply with this rule could result in chatter and inaccuracies in hole size.

12. *Select standard tools* whenever the operating conditions allow. Standard tools are less expensive, readily available, and interchangeable.

13. *Select the largest toolholder shank* the machine tool will allow. This will minimize deflection and reduce the tool overhang ratio.

14. *Select the strongest carbide insert* the workpiece will allow. This will increase overall productivity and lower the actual cost per insert cutting edge.

15. *Use negative rake insert geometry* whenever the workpiece or the machine tool will allow for it. This will double the cutting edges, provide greater strength to the insert, and dissipate the heat.

16. *Select the largest insert nose radius*, but the smallest insert size, that either the workpiece or machine tool will allow. The smallest size insert will be less expensive; using the largest insert nose radius will improve finish, dissipate heat, and provide greater strength.

17. *Select the largest depth of cut and the highest feed rate* that either the workpiece or machine tool will permit when using carbide inserts. This will improve overall productivity and have a minimal effect on tool life.

18. *Know the workpiece material and its hardness.* It is essential to have a thorough knowledge of the material's machining characteristics. If little is known, start at the lowest given cutting speed for that particular material and gradually increase the speed until optimum results are obtained.

19. *Select the cutting speed in relation to the physical properties of the workpiece.* The **Brinell hardness** number (Bhn) usually gives a good indication of the degree of machining difficulty. (The Brinell hardness number is determined by pressing a 10 mil-

limeter hardened steel ball under a load of 6,600 lbs. into the surface of the part sample. The Bhn determines the hardness of the metal. Harder metals have higher Bhn numbers.)

20. *Increase cutter life by using lower cutting speeds and increasing the feed rate* to the limits allowed by the results desired, the setup rigidity, and the strength of the tool.

21. *For maximum cutter life, the feed should be as high as possible.* Doubling the feed (measured as **chip per tooth**—*the amount of material which should be removed by each tooth of the cutter as it revolves and advances into the work*) will double the stock removal per unit of time without appreciably decreasing cutter life.

22. *Use low cutting speeds for long cutter life.* However, soft, low-alloy materials can be machined at high cutting speeds without seriously affecting cutter life.

23. *Excessive cutting speeds generate excessive heat, resulting in shorter cutter life.* Chip and cutter tooth discoloration are good indications of excessive cutting speeds.

> **"Tech Talk"**
>
> Diamond-coated carbide tools (DCC) compared to conventional carbide tools offer up to 50 times the wear life for about 10 times the cost.

24. *Be aware that when the tool's cutting edges quickly become dull*, without chip or tooth discoloration, either the workpiece is very abrasive or there is a high resistance to chip separation. The cutting speed must be reduced.

25. *Use a sharp corner milling cutter only when the job calls for milling to a sharp corner.* Otherwise, use a cutter with a large corner chamfer. Greater stock removal with lower horsepower requirements will result.

26. *Climb milling will allow higher cutting speeds.* In addition, it will improve finish and lengthen cutter life.

27. *Coarse tooth end mills are preferred for roughing cuts.* Although some operators prefer fine tooth end mills for finishing cuts, it is possible to obtain finishes by increasing the cutting speed and decreasing the feed (chip load per tooth) of the coarse tooth end mill. Thus, the same end mill can be used for roughing and finishing.

28. *Direct cutting forces against the solid portion of both the machine and the fixture.* If work is held in a vise, direct the cutting forces against the solid jaw.

29. *Use coolants to get maximum cutter life* and to permit operating at higher cutting speeds. While coolant is not normally used when milling cast iron, by applying a jet of air as a coolant, the finish can be improved and cutter life can be lengthened.

30. *Fixture design should be simple* and standard components should be used whenever possible.

31. *The fixture and part should be located positively.* Rough, nonflat parts should be supported in three places and located on tooling holes, if possible.

32. *Fixtures and parts should be readily accessible and moveable* during any part of the machining operation and replaceable in exactly the same position.

33. *Fixture design should be simple and foolproof* for part loading to fixture and fixture loading to table.

34. *Always keep clamps close to fixture supports* in fixture design, and consider safety when designing for load and unload capabilities.

35. *Spindle toolholder tapers should be kept clean and checked for nicks and gouges prior to loading in the tool matrix.* They should be properly stored or reconditioned when required.

36. *Care must be exercised on lathe toolholders when changing inserts and cleaning the insert pocket.* Worn or damaged lathe toolholders can void insert precision and cause part inaccuracies. Replace toolholders when necessary.

RECALLING THE FACTS

1. Always choose the _____ drill length that will permit drilling the hole to the desired depth.

2. _____ is the distance in either surface feet or surface meters that a point on the circumference of a tool travels in one minute.

3. Holes to be tapped should be _____, _____, _____, and _____ prior to tapping.
4. Select the _____ insert nose radius but the _____ insert size that either the workpiece or the machine tool will allow.
5. The _____ determines the hardness of the material; _____ metals have higher _____ numbers.
6. _____ is the amount of material which should be removed by each _____ of the cutter as it revolves and advances into the workpiece.
7. Care must be exercised on lathe toolholders when changing _____ and cleaning the insert _____.

SECTION 5
REVIEW

KEY POINTS

- The only way a CNC machine can be efficiently and effectively used is through proper use and care of cutting tools and workholding devices.
- More tool usage can be expected with CNC machines; as a result, tools will be used up three times faster than with manual machines.
- Always choose the shortest drill length that will permit drilling the hole to the desired depth; short, stubby drills run truer, allow the fastest feeds, and improve tool life.
- Tapping is one of the most difficult machining operations because of the problems of chip clearance and adequate lubrication at the cutting edge of the tap.
- The reamer is a multibladed cutting tool, designed to enlarge and finish a hole, mounted on a flexible shank; it cannot correct errors in hole location, hole crookedness, etc.
- Boring is the process to correct errors in hole location and crookedness. For the best rigidity and accuracy from boring operations, the boring bar should be as short as possible.

Wow! There's a lot to know about tooling and fixturing for CNC machines! We'd better review!

- With the exception of drills, the most widely used, efficient, and economical metal-removal tool for a machining center is the end mill.
- Carbide inserts are available in a range of grades, types, and sizes; generally, the smaller the inscribed circle (IC) of a carbide insert, the less it costs.
- Large radius inserts leave a better surface finish than small radius inserts and have a stronger cutting edge to throw off heat better.
- Heat is the primary cause of insert wear and destroyer of insert life. Other sources of insert wear include: cratering, chipping, too high or too low feeds and speeds, too much of a depth of cut, and various insert problems caused by tool pressure and stress.
- Coolant usage is an important factor related to cutting tool life because it cools and carries heat from the cutting edge of the tool and workpiece, lubricates and reduces heat and friction, and "dampens down" fine dust and particles that are a safety and health hazard to the operator.
- The most important thing to be aware of about toolholders is that they be free from nicks, gouges, dirt, grit, and any other visible signs of damage.
- A jig is a part-holding and -locating device that also positions and guides the cutting tool; a fixture only holds and locates the workpiece.
- Proper fixturing is very important to successful CNC machining; a poorly designed or manufactured fixture often causes problems at the machine, holds up production, or produces much waste.
- Modular fixturing makes use of reusable components that can be assembled and torn down based on changing part requirements.
- Safety and rigidity are the most important items to be considered when locating and securing parts for machining operations. Good setup practices begin with the use and care of proper clamping and workholding surfaces.
- An unsecured or nonrigid setup can cause damage to the holding fixture, cutting tool, and machine in addition to a scrapped

part. A well-secured and rigid setup reflects the operator's mechanical abilities and good judgment.
- Regardless of whether a manual, N/C, or CNC machine is to be used to cut metal, it is very important to follow sound tooling practices.

REVIEW QUESTIONS

1. Explain, in your own words, why tooling considerations are so important to success on a CNC machine.
2. What is the most important factor to consider when selecting a twist drill for a CNC machine?
3. Multiple-diameter, multiple-land drills are sometimes called _____ drills.
4. Generally, taps are divided into which two major classifications?
   a. straight flute and spiral flute
   b. fluted and fluteless
   c. hand taps and machine screw taps
   d. conventional taps and bottoming taps
5. What are the most common causes of tap breakage on CNC machines? How can they be avoided?
6. What is meant by the *length-to-diameter ratio* (L/D) in boring?
7. Carbide insert size is measured by the largest circle that will fit entirely on the insert, called the _____.
8. The primary cause of insert wear and destroyer of insert life is:
   a. too high or too low feeds and speeds
   b. cratering and chipping
   c. wrong grade selection
   d. heat
9. What are the advantages of using cutting fluids in machining applications?
10. What is the difference between a *jig* and a *fixture*?
11. Why is fixturing so important to the success of a CNC machine?

12. What is meant by *foolproofing* a fixture?
13. List the advantages of modular fixturing over hard or conventional fixturing.
14. Safety and _____ are the most important items to consider when locating and securing parts for machining operations.
15. List five general clamping guidelines for CNC machines.
16. List and explain five new words you learned in this section.
17. Using full sentences and in your own words, write a paragraph about the value and importance of modular fixturing.

SHOP ACTIVITIES

1. Go out in the shop to locate and identify different types of:
   a. *fixtures.* Notice the placement of clamps, locators, and how the part is positioned and clamped in the fixture. How is the fixture or fixtures oriented to the CNC machine table?
   b. *inserts (old and new), cutting tools, and toolholders.* Notice wear on the old or discarded inserts. Do you see any evidence of cratering, chipping, or cracking? What kind of shape are the toolholders in? How are conventional twist drills, end mills, etc. reconditioned?
   c. *clamps.* Identify the different types along with individual locators, pushers, jacks, and clamping bolts. How are most of the setups made in your shop? How is the setup information communicated to machine operators? How and where are clamps and tooling stored? Are they stored in a general and easy-to-access spot for all operators?

   Set up a class or a discussion group to discuss these topics and tooling in general. Make suggestions for improvement and present them to your instructor or supervisor.

2. Call local tooling/carbide insert vendors. Set up a time for them to come in to your school or shop (with samples) to discuss inserts, holders, grades, and coatings of carbide. They should also be prepared to discuss machining of different types of material with their tooling. Most are eager and willing to put on presentations.

SECTION 6

ADVANCED CNC CAPABILITIES

OBJECTIVES

After studying this section, you will be able to:

- Describe new CNC features and advanced applications.
- Explain past and present CAD/CAM application capabilities.
- Identify and explain the four types of manufacturing cells.
- Discuss language- and graphics-based computer part programming.
- Describe four new automation concepts and capabilities.

I hear a lot of people talking about advanced CNC, graphics, CAD/CAM, cells, FMS, and automation. What's it all about anyway?

INTRODUCTION

At this point in your studies of computer numerical control, you have had a good introduction as to what CNC is, what it does, and how it works. You have also had an opportunity to study how simple part programming is done for a CNC lathe and machining center. However, more advanced CNC and machine tool capabilities exist than what have been briefly covered so far in this text. In addition, new and advanced computer programming methods for CNC machines have been accepted and are in wide use.

Beginning with an introduction to language- and graphics-based computer programming, this section will deepen and broaden your exposure to CAD/CAM applications. In addition, new CNC and machine tool features, functions, and capabilities are discussed followed by a discussion of flexible manufacturing cells and systems and automation concepts and applications.

Hopefully, this section will provide you with an overview of the new, interesting, and exciting things happening in computer applications and CNC and you will be encouraged to further your studies in this field.

UNIT 32

New CNC Features and Advanced Applications

KEY TERMS

shop floor programming (SFP)
composites
waterjet machining
part features
central processing unit (CPU)

ADVANCEMENTS IN CNC MACHINE TOOL PERFORMANCE

What began as an idea to control machine tools by numbers in the early 1950s by John Parsons has developed into a manufacturing concept that has changed the metalworking industry forever. New CNC features, functions, capabilities, and advanced applications are continuously being researched, developed, and improved.

The control systems of today do essentially the same things that they did 15 years ago, move slides around and control the spindle and other action. But major improvements in both software and hardware have occurred in the last few years. Many of the new CNC operational characteristics, user-friendliness, and programming advancements are all a result of the software improvements. The cost of the new software development can be as much as 70 percent, or more, of the control.

CNC hardware (Figure 32–1) has also advanced considerably, primarily centered around the development and adoption of the 32-bit microprocessor. This has dramatically improved processing speed, in many cases up to 10 times faster than the older version, the 16-bit control. A very good 16-bit CNC will process a block of new coordinate values in three axes in 50 to 60 milliseconds (thousandths of a second). A 32-bit processor can process a block in as little as two milliseconds.

Other features of typical 32-bit processors in CNCs include:

- 75,000 ft. of part program storage

FIGURE 32–1
The development of the 32-bit microprocessor has increased the processing speed of CNCs 10 times over the older, 16-bit version. (Courtesy of Cincinnati Milacron, Inc.)

- 20-megabyte hard disk within the control (one megabyte equals one million bytes of data)
- improved problem diagnostics
- tool and part management
- better communications capabilities
- multi-tasking (doing more than one task at the same time) operating system which permits the machine tool to operate at maximum capability while the operator performs other CNC tasks such as program downloading and tool data updates
- improved graphics; allows the programmer to see the outcome of his/her work on the CRT screen before any metal is cut or tools broken

Microprocessors, once used sparingly in CNCs, are increasing in new capabilities and tasks to be performed at the shop floor level, while decreasing in size and cost. Additionally, they are simpler and easier to use, with much higher reliability. Collectively, the parts of a CNC that work together to provide improved CNC machine tool performance include:

1. *central processing unit* (CPU)
2. electro/mechanical servo control systems for spindle and axis control
3. various machine, human, and auxiliary interfaces
4. the internal bus structure, or information highway, which passes information from one portion of the control system to another

MODERN PART PROGRAMMING TECHNOLOGY

One new CNC feature that is available today is called *shop floor programming,* or *SFP* for short. SFP (sometimes referred to as *conversational programming*) is an automated part programming system embedded in a CNC with powerful capabilities equal to many commercially available off-line personal computer-based programming systems. Unlike the hard-wired N/C versions of the 1950s and 1960s, CNC in the 1990s does not require significant training. Programming input can be in conversational English. That is, you do not need to know all the G codes, M codes, and so forth you learned earlier. To drill, for example, you just need to input the word "drill" or choose it from a menu and the positions where you want to drill.

The shop floor programming feature of a CNC makes it very easy to program directly from a part drawing. Even a relatively inexperienced programmer can obtain quick and accurate results by inputting part dimensional data directly from the part print and using conversational English to direct the machine tool. SFP provides the capability to run in background mode and create a new program or modify an existing program right at the machine while the machine is busy cutting a previously programmed part. The operator/programmer can enter the material name and part-shape dimensions and the system will automatically determine tool types, speeds, feeds, and cutting tool order. In the programming mode, the CRT prompts the programmer through the entire programming process and displays the results. Then it produces a complete program. Cutter path instructions are generated along with all of the preparatory and miscellaneous codes, functions, and commands—all invisible to the user. After any necessary changes or edits are performed, the programming system graphically displays the machining process on the CRT screen (Figure 32–2) while checking for any possible tool or part interferences within the program.

Some CNC manufacturers offer SFP as an optional feature, like buying an option on a new car, while others offer SFP as a standard part of their CNC package. On some SFP systems, the display may be only the path plot of the cutter's centerline, while on other systems there may be a very detailed graphics display showing an image of the part, the cutter path, the cutter, and even the simulation of stock removal. SFP should not be confused with manual data input (MDI), which you have learned is strictly a manual means of in-

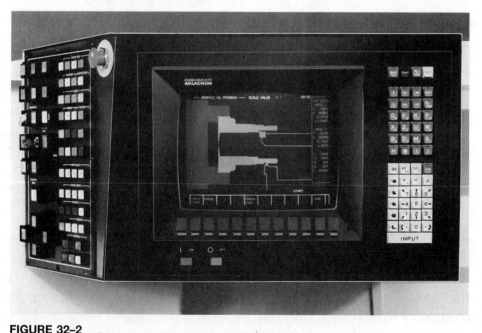

FIGURE 32–2
Modern CNCs graphically display the machining process on the CRT screen and check for any possible tool or part interferences in the program. (Courtesy of Cincinnati Milacron, Inc.)

putting block-formatted data (G codes, M codes, etc.) and is a standard feature on all CNCs.

Other improvements for CNC controllers are taking place in the area of high-level user graphics, more powerful microprocessors, additional memory, and advanced communication systems. These new systems will be capable of accessing engineering data directly from an engineering database, whether in the same building or thousands of miles away, downloading engineering geometry directly to the MCU for part processing and bypassing off-line programming entirely. Part programming as we know it today will become increasingly automated, centered around *part features* such as holes, slots, contours, threads, etc.

COMPOSITES

Advanced applications of CNC continue to expand into areas other than the more traditional metal-cutting, laser drilling, and forming. *Composites* manufacturing is one area being enhanced through the use of improved machine and CNC technology.

The use of composites started in the aerospace industry because of the high strength-to-weight characteristics of composite materials. Kevlar, graphite, or boron fibers combined with a bonding matrix, such as epoxy, provide lightweight structures that are stronger and can flex better than aluminum. Composite parts are used in many military and commercial airplanes today. The "laying up" or forming of composite parts is a rigorous process of building up, layer upon layer, the composite material to construct the general part shape and thickness. Applying pressure, heating, and curing the layered fiber resin material brings the composite material to a finished rigid state.

Machines employing the use of this advanced technology are multiaxis, tape-laying, CNC gantry machines similar to the one depicted in Figure 32–3. The tape-laying head, shown in Figure 32–4, acts like a giant Scotch tape dispenser, precisely laying ply-on-ply of very thin (.005 to .014 inches thick) tape at speeds up to 100 feet per minute. This includes automatic dispensing of the tape from a large diameter supply roll, placing it directly on the underlying mold, applying the proper compaction pressure to seat and debulk, taking up and storing the backing paper, and, at the end of each course,

FIGURE 32–3
An advanced CNC gantry tape-laying machine. (Courtesy of Cincinnati Milacron, Inc.)

FIGURE 32–4
A tape-laying head, acts like a giant Scotch tape dispenser, laying ply-on-ply of very thin composite material at speeds up to 100 feet per minute. (Courtesy of Cincinnati Milacron, Inc.)

cutting the tape (but not the backing paper) to the required length and angle.

WATERJET MACHINING

Another recent and important CNC application is the development of the waterjet and abrasive *waterjet machining*. Today, there is no American airplane, car, or house that does not have something in it that was cut with a waterjet or an abrasive waterjet.

A waterjet or abrasive waterjet (Figures 32–5 and 32–6) is a single-point sandblaster, so to speak, which may be pointed in almost any direction, does not wear out, is very safe to use, and causes very little, if any, damage to the material being cut. After the material is cut, the water or water-abrasive stream is collected in a tank or reservoir.

Waterjet machining can cut a wide variety of metal and nonmetal materials using a small diameter (0.004–0.024 in.) focused stream of water passing through a small hole under high pressure (20,000–60,000 psi, or pounds per square inch) traveling at high velocity

FIGURE 32–5
A CNC waterjet machine is essentially a single-point sandblaster that does not wear out, is very safe to use, and causes little if any damage to the material being cut. (Courtesy of Jet Edge.)

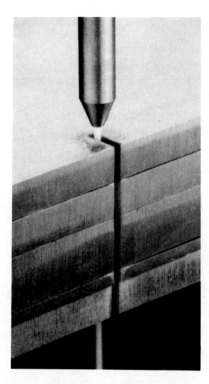

FIGURE 32–6
Waterjet machining can cut a wide variety of metal and nonmetal materials by focusing a stream of water through a small hole using high-pressure and high speed. (Courtesy of Jet Edge.)

(1,700 to 3,000 ft/sec). When the force of the waterjet impacting the surface exceeds the strength of the material, the material is cut.

Using this method, materials can be cut in multiple directions without ragged edges (unless the traverse speed is too high), without heat, and generally faster than a bandsaw. Waterjet machining produces no heat problems nor does it cause the material being cut to come apart or be deformed. Waterjet and abrasive waterjets can best handle materials that gum up saws or suffer unacceptable heat damage from other processes. Plasma and lasers, for example, leave a heat-affected zone on the material they cut.

Sometimes referred to as *hydrodynamic machining*, waterjet machining was improved further by the addition of abrasives, such as silica and garnet, into the stream to cut metals, composites, and other hard materials. CNC waterjet and abrasive waterjet machining continues to be developed and accepted more among manufacturers.

RECALLING THE FACTS

1. A 32-bit processor in today's CNC includes 75,000 ft. of part program storage, _____ hard disk, improved problem diagnostics, _____, better communications capabilities, _____ operating systems, and _____.

2. The parts of a CNC that work together to provide improved machine tool performance are:

   a. _____

   b. electro/mechanical servo control systems for _____ and _____ control

   c. various machine, human, and auxiliary _____

   d. the internal bus structure, or _____ _____, that passes information from one part of the control system to another.

3. _____ is an automated part programming system embedded in a CNC that allows input to be in conversational English.

4. _____ should not be confused with manual data input or _____, which is strictly a manual means of inputting block-formatted data (G codes, M codes, etc.)

5. In the future, part programming as we know it today will become increasingly automated and centered around _____ such as holes, slots, contours, threads, etc.

6. The use of _____ materials started in the aerospace industry because of the high _____-to-_____ characteristics of _____ materials.

7. A _____ or abrasive _____ is a single-pointed sandblaster, so to speak, which may be pointed in almost any direction, does not wear out, is very safe to use, and causes little, if any, damage to the material being cut.

8. Sometimes referred to as *hydrodynamic machining*, _____ has been improved further by the addition of _____, such as silica and garnet, into the streams to cut various materials.

UNIT 33

Language- and Graphics-Based Computer Programming

KEY TERMS

language-based systems
automatic programmed tool (APT)
postprocessor

interactive system
workstations
CL data
CAD/CAM

graphics-based system
mainframe computer
personal computers (PCs)

LANGUAGE-BASED PROGRAMMING

Using the computer to prepare programs for numerically controlled machines began in the early 1960s with *language-based systems*. As you remember, N/C was developed as an answer to some complex aerospace machining problems, such as the curves of blade and airfoil surfaces. Language-based computer programming systems developed the same way, through a need to machine complex surfaces with a simplified English-like processor language.

At that time, many aerospace companies tried to write their own processor language. However, the companies found that the job required more time and effort than could be committed. Finally, the members of the AIA (Aerospace Industries Association) pooled their resources in a cooperative development project. In 1961, it was decided to broaden their scope. Further development was turned over to IIT (Illinois Institute of Technology) research in Chicago. The *APT (automatic programmed tool)* language-based system soon emerged, representing over 100 years of development and testing.

Manufacturers soon found that by using a computer and a language-based processor to prepare N/C programs, the time and cost of tape preparation was greatly reduced, particularly if the part was complex. It was also found that processor-based languages produced accurate programs more often than manual programming and could provide error messages when the English-like words were not input correctly, were misspelled, or certain types of technical problems occurred. Making the corrections and running the program through the computer again permitted corrections to be made at the computer level, where it is generally cheaper to correct, rather than finding mistakes and errors at the machine tool. This reduces the amount of machine time that might normally be wasted because of tape or programming errors. Figure 33–1 illustrates the flow of an N/C program when using a computer with a language-based processor.

Over time, new and different languages began to be developed, including multiple versions of APT itself. Some of these other language-based programming systems exist today, but APT is still the most common and widely used language-based processor for N/C applications.

With APT or other language-based programming systems, you define the part in terms of geometry and then, through tool motion statements, direct the cutter positioning or routing around the defined geometry. An example of a simple part and a simple APT program is seen in Figure 33–2. As you can see from the example, APT is a language that is somewhat English-like but has its own vocabulary and must be learned like any other language, such as French, Spanish, or German.

APT part programming involves three major elements:

1. Defining and naming of the part's geometric points and surfaces.
2. Specifying the cutting tool and action or tool motion statements. (These statements move the cutter to the points or along the defined geometric surfaces.)
3. Specifying the conditions required at the machine tool, such as spindle speeds, feed rates, and other function commands.

The output from APT is called centerline *cutter location,* or **CL data,** for short. The APT tool centerline data must be converted into the machine language code that a machine tool controller (MCU) can understand. You have already seen machine language code. It con-

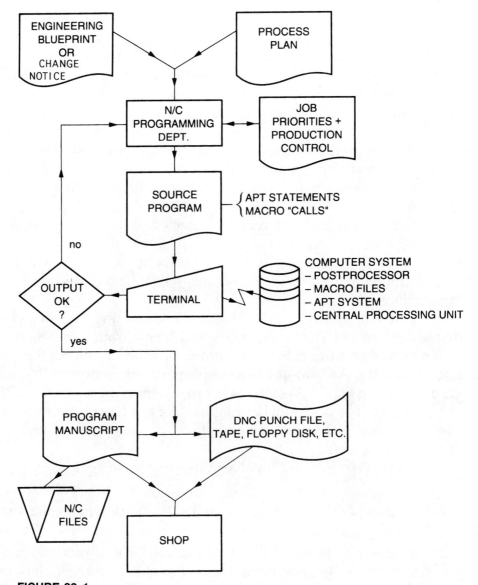

FIGURE 33–1
Flow of an N/C program when a using a computer with a language-based processor.

sists of the X, Y, Z, G, and M words, and so forth. This conversion is done by a *postprocessor*. A postprocessor is a set of computer instructions which transforms tool centerline data into machine motion commands. Essentially, each different combination of machine tool/control unit requires its own postprocessor. Whether a lan-

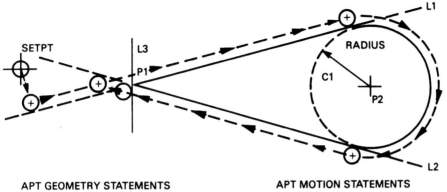

APT GEOMETRY STATEMENTS
SETPT = POINT/X, Y, Z
  P1 = POINT/X, Y, Z
  P2 = POINT/X, Y, Z
  C1 = CIRCLE/CENTER, P2, RADIUS, R
  L1 = LINE/P1, LEFT, TANTO, C1
  L2 = LINE/P1, RIGHT, TANTO, C1
  L3 = LINE/P1, ATANGL, 90
  CUTTER/.5

APT MOTION STATEMENTS
FROM/ SETPT
RAPID
GO/ TO, L1
FEDRAT/20
TLLFT, GOLFT/L1, TANTO, C1
GOFWD/C1, TANTO, L2
GOFWD/L2, PAST, L3
GOTO/SETPT
FINI

FIGURE 33–2
A simple part and a simple APT part program.

guage-based processor or a graphics-based system is used to generate the part program, a postprocessor program must still be written for each different machine tool/control unit combination. This important relationship is illustrated in Figure 33–3.

The APT geometry statement consists primarily of three parts:

$$\underset{\text{Symbol}}{(1)} = \underset{\text{surface}}{(2)} / \underset{\text{description}}{(3)}$$

Example:

$$C1 = Circle/center, P2, radius, R$$

The first part is a symbol, which is a name assigned to a particular geometric element (in this case it is C1). This symbol is then set equal to the definition (the second part of the statement) which is a major word such as *point, line, circle*, etc. The major word defines the type of surface or geometric element that the symbol represents. The third part of the APT geometry statement is the actual description, which consists of minor words or modifiers (using the APT vocab-

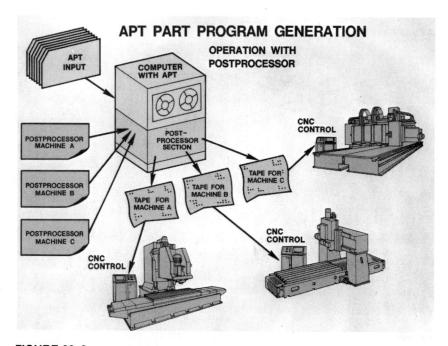

FIGURE 33–3
The APT processor, post processor, and machine tool relationship. (Courtesy of Cincinnati Milacron, Inc.)

ulary words) and numerical values of the point, line, circle, etc. These position the element in space and determine its specific size.

Motion statements in the APT language are typical of statements that might be used in directing a person to walk around the block or through town (go left, go right, left side, right side). English-like APT vocabulary words such as TLLFT (tool left), TLRGT (tool right), GOLFT (go left), and GORGT (go right) are used to direct the tool path relative to the defined geometry.

The following is an example of an APT tool motion statement:

TLLFT, GOLFT/L1, TANTO, C1

A simple example of another APT program is shown in Figure 33–4.

GRAPHICS-BASED PROGRAMMING

In the late 1970s and early 1980s, a new term called *CAD/CAM* was born. CAD stands for computer-aided design and CAM stands for

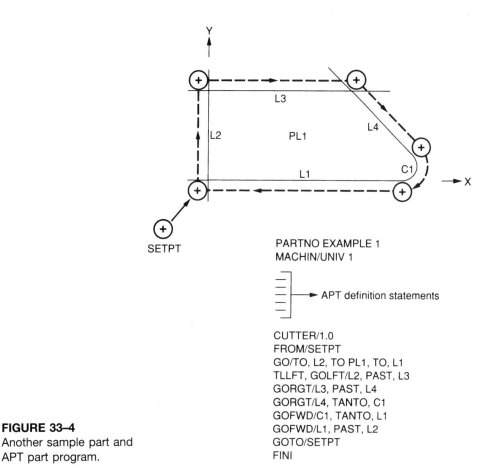

FIGURE 33–4
Another sample part and APT part program.

```
PARTNO EXAMPLE 1
MACHIN/UNIV 1

    APT definition statements

CUTTER/1.0
FROM/SETPT
GO/TO, L2, TO PL1, TO, L1
TLLFT, GOLFT/L2, PAST, L3
GORGT/L3, PAST, L4
GORGT/L4, TANTO, C1
GOFWD/C1, TANTO, L1
GOFWD/L1, PAST, L2
GOTO/SETPT
FINI
```

computer-aided manufacturing. These systems use the computer to do graphics part and assembly design and analysis, as well as N/C manufacturing. The early CAD/CAM systems were primarily mainframe computer-based. *Mainframe computers* are large multipurpose computers that link several different terminals and users to the system at the same time. One mainframe computer can support many CAD/CAM terminals in addition to other business and technical applications. As integrated circuits, advanced microprocessors, and other technologies advanced, stand-alone (allowing only one person to use at a time) computing platforms, called *workstations* and *personal computers (PCs)*, were developed. Over time, the workstations and PCs became the dominant computing platform because they were now capable of mainframe power on a PC or workstation and because they could be purchased for a fraction of the mainframe cost. At the same time, powerful CAD/CAM software was developed to run on the workstations and PCs.

CAD/CAM systems allow for detail parts and assemblies to be designed with part geometry being stored electronically in a central database. Storing data electronically allows for easy access and retrieval of part data as opposed to storing actual drawings.

CAD/CAM systems are designed to be user-friendly and prompt the user to do things through a menu selection display, as seen in Figures 33–5 and 33–6. In many cases, they use construction techniques familiar to the conventionally trained draftsman. This greatly helps a designer's productivity, creativity, and conceptual thinking. Before CAD, the designer would sit in front of a drawing board with paper and pencil; now that person sits in front of a graphics terminal or workstation and creates geometric images on the screen.

The designer can also add and reproduce dimensions and symbols as well as move and rotate the constructed images in a variety of ways never before possible with conventional paper and pencil methods. Once the engineer/designer has arrived at the final version or design, the image, because it is based on mathematical coordinates and values electronically stored within the computer, can be transmitted to other computer-related devices for printing, plotting, etc.

FIGURE 33–5
A CAD system in use. (Courtesy of International Business Machines Corporation.)

LANGUAGE- AND GRAPHICS-BASED COMPUTER PROGRAMMING • 327

FIGURE 33–6
CAD systems are designed to be user-friendly and prompt the user to do things through pull-down menus. (Courtesy of International Business Machines Corporation.)

Although some language-based N/C programming systems, primarily APT, are still presently in use, graphics-based N/C systems are state-of-the-art today. Language-based systems provided a big productivity boost over manual programming methods but their continued used and development presented several distinct problem areas:

1. Detailed part geometry, created electronically using CAD systems and stored in an engineering computer database, had to be re-created using APT (APT geometry statements) or other language definition statements. This is doing double work. Why re-create part geometry that already electronically exists in the database?

2. Language-based processors are text only—no graphics. Producing an accurate program depends on the programmer's ability to "visualize" the machining process in his or her mind while writing geometry and tool motion statements. With graphics-based systems, verifying and checking for tool path

accuracy, fixture interferences, and their machining related problems can be accomplished visually by watching the CRT screen. With APT, these checks can only be accomplished through close mathematical checking of the postprocessor and CL output, plotting, and tape prove-out on the machine.

3. There is a declining base of skilled, trained programmers knowledgeable in APT and other processor languages.
4. Language-based programming is entirely too time-consuming and error-prone. Manufacturers must reduce the overall product development, manufacturing, and assembly cycle time in order to bring products to market faster.

Interactive Systems

Graphics-based systems were introduced to greatly simplify the programming process, move visualization from the programmer's mind to the computer screen, and provide the opportunity for immediate feedback to the programmer. These systems are called *interactive systems,* which means they allow the user and the computer terminal to interact (the system provides both input and output similar to the way you play video games).

Interactive systems, as seen in Figure 33–7, make use of color graphic, multi-tasking, pop-up menu displays that can design a part in three dimensions, create detailed part prints, design tools and fixtures, and write the N/C program to produce the part—all with the same system. Graphics-based systems, many of which are PC-based, use a mouse as a pointing device (Figure 33–8) along with easy-to-understand menu guides to enter and change geometry and tool motion. Users can easily define part geometry, create the tool path, reduce errors by graphically seeing the results, and greatly improve productivity through the interactive capability to reprogram and replot the part. Also, the majority of N/C prove-out can be accomplished on the graphics display terminal rather than tying up the CNC machine and operator on the shop floor.

The CAD-like "front end" of N/C programming systems permits users to create a part by drawing lines, circles, arcs, and planes. Interactive graphics helps N/C programmers create the tool path by defining the exact path the tool will take to cut the part. N/C programming using graphics enhances the part programmer's ability

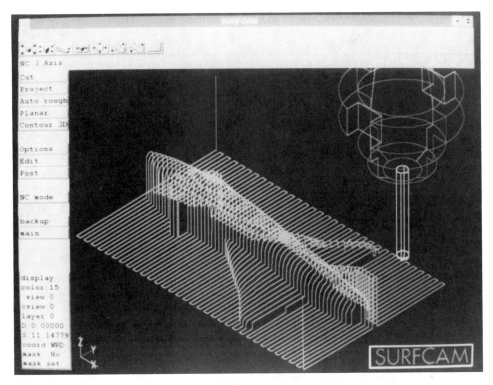

FIGURE 33–7
Graphics-based systems make use of color graphic, multitasking, pop-up menu displays that can perform a variety of CAD/CAM functions. (Courtesy of Software Inc.)

FIGURE 33–8
Graphics-based systems, many of which are PC-based, use a mouse as a pointing device and are very easy to enter and change geometry and tool motion. (Courtesy of International Business Machines Corporation.)

to visually follow the tool path by obtaining a three-dimensional view of cutter clearance planes, retract planes, depth planes and clamps, fixtures, and casting clearances.

Tool path editing capability, now available in some N/C packages, provides the N/C programmer with an additional and practical technique to modify tool paths that have been generated. Calculated tool paths are then verified and the computed coordinates are transmitted to a postprocessor to generate the N/C output. Thus, an N/C programmer can sit in front of a graphics display terminal, just like the designer, create the tool paths and then watch the calculated and verified tool paths occur in relation to the actual part.

RECALLING THE FACTS

1. A _____-based system contains a _____ that accepts an English-like input language for programming complex parts.
2. Manufacturers found that by using a computer and a _____ _____ _____ to prepare N/C programs, the time and cost to prepare those programs could be greatly _____.
3. _____ stands for *automatic programmed tool;* today, _____ is still the most widely used _____-based _____ used for N/C applications.
4. Using the _____ computer part-programming language requires:
   a. defining the _____,
   b. specifying the tool _____, and
   c. specifying the conditions at the machine tool, such as _____ and _____.
5. The output from APT is called _____ data, which is the _____ _____, or _____ data.
6. A _____ is a set of computer instructions which takes the output from APT and converts it into the machine code (Gs, Xs, Ys, Ms, Zs, etc.) that the MCU can understand.
7. _____ stands for computer-aided design and _____ stands for computer-aided manufacturing.

8. _____ computers are large multipurpose computers that link several different terminals and users to the system at the same time. As integrated circuits, advanced microprocessors, and other technologies advanced, stand-alone computing platforms called _____ and _____ were developed.
9. _____-based systems were introduced to greatly simplify the programming process, move _____ from the programmer's mind to the computer screen, and provide the opportunity for immediate _____ to the programmer.
10. An _____ system means it allows the programmer and the computer to _____, just like in video games, providing both _____ and output.

UNIT 34

CAD/CAM Basics and Beyond

KEY TERMS

CAD	CAM	surface model
database	associativity	wire-frames
Initial Graphics Exchange Standard (IGES)	primitives	

BASICS OF CAD/CAM

What is a CAD/CAM system? As you have already learned, *CAD* stands for computer-aided design and *CAM* stands for computer-aided manufacturing. A CAD system is composed of hardware and software elements. Hardware elements are visible and consist of the computer, the screen display unit or monitor, the keyboard and mouse, and an output device like a printer. Software elements are invisible and consist of the application programs written to draw, color, rotate, and join the geometric figures and objects you create on the screen.

The simplest operations that can be performed on the screen of a CAD system are drawing points, lines, circles, and arcs. Splines or French curves can also be generated, a capability that is essential for the creation of complex surfaces. Solids are created and built by adding and subtracting the intersections of solid objects called *primitives,* such as cylinders, cones, and cubes.

CAD systems allow for detail parts and assemblies to be designed in an interactive environment, with design geometry being stored in a central database for access and retrieval. An interactive system allows the user to interact with the system by providing immediate responses to the user's instructions, changes, or additions through a screen-driven menu selection, as seen in Figure 34–1. This greatly enhances a designer's productivity, creativity, and conceptual thinking, as you learned in Unit 33.

CAD/CAM BASICS AND BEYOND • 333

FIGURE 34-1
Interactive systems allow interaction with the system by providing immediate responses to the user's instructions. (Courtesy of International Business Machines Corporation.)

The simplest of CAD systems are two-dimensional (2-D) (just like part prints and pencil-and-paper engineering drawings). The vast majority of CAD usage is 2-D. CAD systems are primarily used to design detail parts and assemblies through the creation of lines, surfaces, solids, intersections, and curved surfaces. In simpler terms, they create, transform, and display geometric figures through the use of computer hardware and software.

As CAD/CAM systems began to make a big impact on the engineering/manufacturing process in the mid- to late 70s and early 80s, hardware elements underwent considerable change and cost reductions. At the same time, software increased in both capability and functionality. The overall result was mainframe power on a PC or workstation for a fraction of the mainframe cost. Graphics terminals have now become smaller and more powerful, color displays have become standard, and very powerful programming techniques for manipulating computerized images have been developed.

DESIGNING WITH THE CAD SYSTEM

CAD capabilities have advanced rapidly since their beginnings and range from using computers to create drawings to performing isolated calculations and compiling a bill of materials. Recorded im-

> **"Tech Talk"**
>
> Increasingly, in the future, feature-based CNC machining will be used based on a powerful collection of manufacturing and machining software decision logic for specific part features such as holes, slots, contours, threads, etc. These packaged, pre-programmed, pre-proven machining modules will be automatically selected based on design feature data already available via the CAD geometry. Work continues on developing and perfecting feature-based machining logic.

ages that can be electronically stored range from a simple straight line to a multicolored image of a three-dimensional assembly. Some images feature sculptured surfaces and moving parts, with shading and perspective to show depth visualization. These descriptive geometric representations can then be rotated and viewed like an object in space, giving the designer total part viewing capabilities.

You should not be overwhelmed by the complexity of the technology surrounding CAD/CAM systems. You do not need to know the inner workings of a CAD system in order to use one productively any more than you need to know how the engine and transmission of a car work in order to drive one. Using a CAD/CAM or any computer system is a lot like driving a car; once you get to know how it works and how to make it do what you want it to do, along with understanding the "rules of the road," it gets easier and you gain confidence. Remember, everybody who uses a CAD/CAM system today had to start at the beginning at some point in their lives.

In order to make things easier and more productive for the user, CAD systems have a library of stored shapes and commands to help the input of designs. They perform four basic graphic-oriented functions that can enhance the productivity of a designer using a CAD system.

1. CAD systems allow you to copy a part of the design. The ability to take part of an image (a bolt-hole pattern, for instance), copy it, and use it in several other areas of the design, when a product has features that can be repeated throughout the design, is a big productivity advantage.
2. The system can capture and reuse portions of one design file in another. That is, that same bolt-hole pattern, for example, can be copied from the original part and used on other parts that require that same bolt-hole pattern. This is a particularly important and productive feature when changes are being

made to existing designs or if an entire "family" of parts requires the modification.
3. CAD systems have the ability to "zoom in" on a small part of the image, like the zoom lens on a camera, or change the size or proportion of one part of the image in relation to another. This zooming feature improves the quality and productivity of detail drafting.
4. CAD systems have multiple-view capability, which allows the CAD operator to rotate or flip the figure on the screen to see the design from different angles or perspectives.

One of the most important aspects of a CAD/CAM graphics system is that once the final part image is created, it can be stored electronically in the engineering *database*. A database allows you to electronically store and retrieve part geometry. This makes it readily accessible for viewing (Figure 34–2) by engineering personnel and, ultimately, for manufacturing use. By being able to readily access this database, the CAD operator has the option to call up, copy, and reuse a past design in its entirety or modify a past design to meet current requirements. If a printed copy of a part or design is re-

FIGURE 34–2
Engineering drawings are stored in an electronic engineering database. The database allows electronic storage and easy retrieval of part geometry. (Courtesy of International Business Machines Corporation.)

quired, output can be directed and transmitted to printers or plotters connected to the computer.

From a design point of view, the electronic database storage of images greatly enhances and aids compatibility and part interference visualization. Different images can be called up and merged, stacked, and rotated for assembly viewing and clarification without having to draw the assembly on paper and then find some assembly interference factors necessitating redesign. As mentioned previously, replication or duplication of details is also possible. A designer may need to construct details, such as a screw or a bracket, only once and then duplicate and locate it as necessary. These details can then be stored in a library of standard figures or details for other designers to call up and use. Most engineering database systems provide both data management and data protection. That is, the CAD system controls the deletion or erasing of data and protects against unauthorized changes to drawings.

VISUALIZING WITH CAD/CAM

With today's advanced systems, three-dimensional (3-D) surface and solid models can be created. A *surface model* is one in which the surfaces, edges, and primary curves of the model are defined.

FIGURE 34–3
Surface models are those in which the surfaces, edges, and primary curves are defined and appear as a solid object on the screen. (Courtesy of International Business Machines Corporation.)

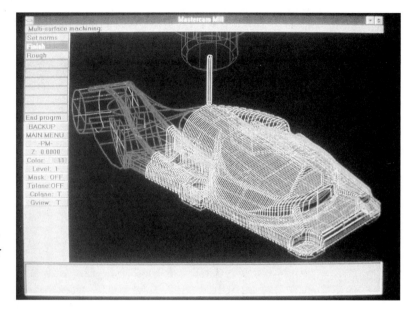

FIGURE 34–4
Wire frame models appear as stick figures where only lines representing edges or intersections are visible. (Photo courtesy of CNC Software, Inc.)

The appearance is that of a solid object, as seen in Figure 34–3. Each location has an X, Y, and Z coordinate and provides a more accurate representation of an object.

Some 3-D models are generated as stick figures of wire-frame objects called *wire-frames* (Figure 34–4) where only the lines representing edges or intersections are visible. The user can see through the wire-frame. With appropriate software, the hidden lines can be removed and surfaces can be stretched over a wire-frame model to provide a surface model. The appearance on the screen is that of a solid object. Surface models are usually used for N/C. A solid model freely describes the edges and faces of the design, plus provides the knowledge of how the edges and faces are connected (topology). Models done in 3-D can be used to produce 2-D drawings.

Graphics creation of a design image is done in three phases. The design or model is first converted from an electronic image to a precise visual form. Second, the display-format image is stored and then mapped to a display. Finally, the image is displayed on the screen. Solid modeling provides the best and most complete representation of an object but extensive workstation or PC computing power is required to handle the workload requirements.

PROGRAMMING WITH CAD/CAM SYSTEMS

CAM is the application of manufacturing to the CAD design. Particularly, this means graphical N/C programming. Basically, the CAM part of a CAD/CAM package accepts the CAD generated part geometry and displays it on the screen. By selecting points on the image displayed on the screen and responding to a series of questions on the screen menu, the manufacturing engineer creates the N/C program. Because the system "reads" the geometry and contains built-in routines for various machining operations, the engineer interacts with the system and no longer has to deal with complex math or language-based processor words and terms. Interactive graphics helps N/C programmers create the tool path by defining the exact path the tool will take to cut the part.

Interactive, graphics-based CAD/CAM systems were introduced to greatly simplify the CAM (N/C programming process) as well as the CAD design process. The graphics capabilities allow visualiza-

FIGURE 34–5
Graphics capabilities allow visualization of a part or tool to move from the mind of the programmer to the computer screen. (Courtesy of International Business Macahines Corporation.)

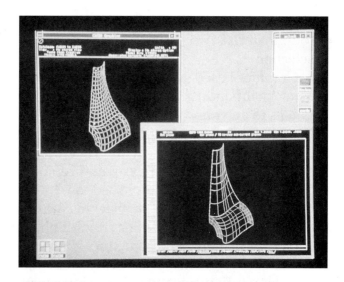

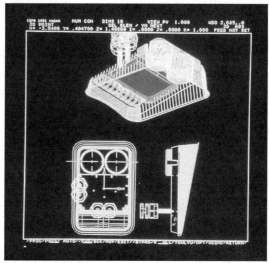

FIGURE 34-6
Many graphics systems today consider complete cutter and toolholder geometry in relation to the workpiece thereby showing tool gouging and interference problems on the screen before the part is actually run on the machine. (Courtesy of International Business Machines Corporation.)

tion to move from the programmer's mind to the computer screen, and provide the opportunity for immediate feedback to the programmer via an interactive system (Figure 34–5). The CAM visualization and built-in computing capabilities permit graphics-based systems to handle design difficulties, such as surfaces that will not blend together, with ease. In addition, many systems consider the complete cutter and cutting toolholder geometry in relation to the workpiece and check-surface geometry (Figure 34–6). This feature allows gouge-free tool paths to be generated. As part of the machining strategy, other features allow menu selection of the machine tool, material type, specific tools for each cut, and speed and feed selection.

Graphics-based systems, many of which are PC-based, use a mouse as a "point-and-click" device along with easy-to-understand menu guides to enter and change geometry and tool motion. Users can easily define part geometry, create the tool paths, reduce errors by visualizing results, and greatly improve productivity through the interactive capability to reprogram and replot. Also, the majority of N/C prove-out can be accomplished on the graphics display terminal rather than idling the N/C machine and operator on the shop floor.

N/C programming using graphics enhances the part programmer's ability to visually follow the tool path by obtaining a three-dimensional view of cutter clearance planes, retract planes, depth planes and clamps, fixtures, and casting clearances. Interactive graphical tool path editing capability, now available in most N/C packages, provides the N/C programmer with additional and practical techniques to modify tool paths that have been previously generated. Calculated tool paths are then verified and the computed coordinates are transmitted to a postprocessor to generate the N/C output. Thus, an N/C programmer can sit in front of a graphics display terminal, just like the designer, create the tool paths and then watch the calculated and verified tool paths occur in relation to the actual part.

As CAD/CAM systems continued to evolve, and because of their proprietary nature, it became obvious that no standards existed for how geometric information was structured and stored. Consequently, geometric data could not be exchanged from one system to another. Early efforts involved creating one-to-one translator programs that would convert one system's proprietary storage structure to another but that proved unmanageable, as multiple translators would be needed for every combination of one-to-one system. Eventually, a standard data format with a neutral database for geometric translation was proposed in 1979 by the U.S. Air Force. This standard data format, known as the *Initial Graphical Exchange Specification,* or *IGES,* was supported by other Defense Department groups and CAD/CAM users and suppliers. IGES became the standard in 1981. Most suppliers now provide an IGES-in and IGES-out capability for their CAD/CAM customers.

One CAD/CAM feature that deserves special consideration is *associativity*. Full associativity between part geometry and interactive machining allows changes to be made to a single geometry model with each tool path automatically changed to reflect the new model. Essentially, when the design model is changed, the N/C tool path is automatically altered to suit the changed geometry. The cutter path, however, must still be regenerated and repostprocessed. This is a tremendous productivity improvement and time saver over conventional language-based programming, where definition and possibly tool motion statements would have to be changed and the program rerun and repostprocessed every time a design engineer changed the part drawing. Additional time is saved because the part geometry changes are now electronic, as opposed to issuing an engineering change notice and new part drawings for every change.

RECALLING THE FACTS

1. The simplest operations that can be performed on a computer-aided _____ or _____ system are drawing points, lines, circles, and arcs.
2. Solids are created by adding and subtracting the intersections of solid objects like cylinders, cones, cubes, etc. called _____.
3. An _____ system allows the user to interact with the system by providing immediate responses to the user's instructions, changes, or additions through a screen-driven menu selection.
4. The simplest of _____ systems are _____-dimensional, just like part prints and paper-and-pencil engineering drawings.
5. One of the most important aspects of a _____ system is that once the final part image is created, it can be stored electronically in the engineering _____.
6. A _____ is one in which the surfaces, edges, and primary surfaces of the model are defined.
7. Some 3-D models are generated as stick figures of objects called _____.

8. Computer-aided _____ or _____ is the application of manufacturing to the CAD design.
9. N/C programming using _____ enhances the part programmer's ability to visually follow the tool path by obtaining a three-dimensional view of cutter clearance planes, retract planes, depth planes and clamps, fixtures, and casting clearances.
10. _____ is a standard data format with a neutral database for geometric translation.
11. Full _____ between part geometry and interactive machining allows changes to be made to a single geometry model with each tool path automatically changed to reflect the new model.

UNIT 35

Flexible Manufacturing Cells and Systems

KEY TERMS

cell
stand-alone CNC machine tool
multimachine cell
manufacturing cell
flexible manufacturing system (FMS)
pallets
single CNC machine cell

In manufacturing terms, a *cell* means arranging or organizing a cluster of manual or CNC machines into logical groups in order to process similar parts or a "family" of parts. Cells can be arranged by part type, machine function, color, part size, or however the manufacturer feels he can best group or arrange his machines to provide the highest productivity. For example, a manufacturer might group all machines necessary to machine shafts together and call it the "shaft cell." He might group all machines necessary to complete all large box-type parts together, paint all the machines blue, and call it the "blue cell."

When companies group machines together into cells, they want to avoid moving parts all around the shop from machine to machine (because this costs time and money) and complete as much of the work as possible in the cell. Today, the term *manufacturing cell* is much broader and, because of the advancements made in CNC and automation technologies, can mean some level of automated part and cutting tool loading, unloading, delivery, or exchange to the clustered machines.

Manufacturing cells can be divided into four general categories:

1. traditional stand-alone CNC machine cells
2. single CNC machine cells, or minicells

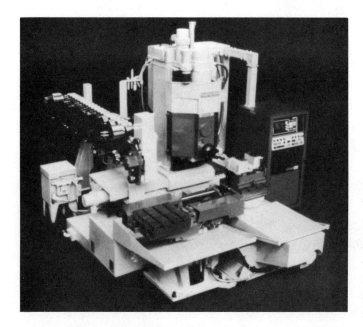

FIGURE 35–1
The stand-alone CNC machine tool is a limited-tool and part storage automatic tool changer and usually functions with one operator running one machine. (Courtesy of Cincinnati Milacron, Inc.)

3. multimachine cells
4. flexible manufacturing systems (FMSs)

The *stand-alone CNC machine tool* (Figure 35–1) is traditionally a limited-tool and part storage automatic tool changer and usually functions with one operator running one machine. Single stand-alone CNC machines have been and still are the type most commonly sold. In many cases, stand-alone CNC machines have been grouped together in a cellular arrangement (Figure 35–2) but still function with one operator running one machine. Stand-alone machines can also be run in a cell of this type with one operator running two or even three machines. Grouping machines together by how many a particular operator is responsible for can further define a cell. Part cycle start and stop times in a cell of this type are controlled and staggered by the operator so that one machine is always running while the other is idled for part loading or unloading.

The *single CNC machine cell,* or *minicell,* generally consists of an automatic work changer with permanently assigned work *pallets* (moveable work tables that can be moved by hand or an automated

FLEXIBLE MANUFACTURING CELLS AND SYSTEMS • **345**

FIGURE 35–2
Stand-alone CNC machines have been grouped together in a cellular arrangement but still function with one operator running one machine. (Courtesy of Cincinnati Milacron, Inc.)

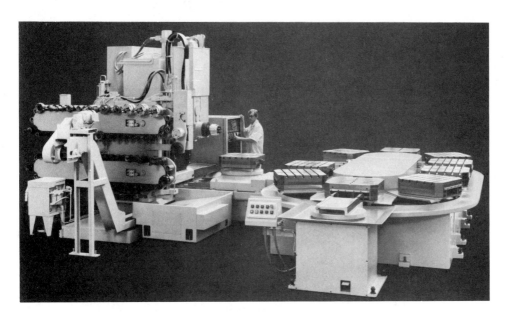

FIGURE 35–3
The single CNC machine cell or minicell consists of an automatic work changer with permanently assigned work pallets or a conveyor robot arm system that is mounted to the front of the machine that allows for additional tool storage. (Courtesy of Cincinnati Milacron, Inc.)

FIGURE 35–4
A multimachine cell designed to load, unload, and progressively move parts from one machine to another. (Courtesy of Cincinnati Milacron, Inc.)

process from machine to machine) (Figure 35–3) or a conveyor robot arm system mounted to the front of the machine. Once loaded and set up with tools and parts, new and special features enable single CNC machines to operate unattended as a self-contained cell. The single CNC machine cell is rapidly gaining popularity because it can be purchased for a fraction of the cost of a complete flexible manufacturing system (FMS).

The *multimachine cell* is made up of multiple metal-cutting machine tools which can be of the same or a different type. Multimachine cells can have an automated method of loading, a progressive method of moving parts from one machine to another, and a method for unloading parts that usually involves using a robot and conveyor system (Figure 35–4) or an in-line system of palletized parts (Figure 35–5). For example, parts can be moved by a robot from a turning center to a center-type grinder and then to an automated gauging system for part completion, all within the cell.

The *flexible manufacturing system (FMS)*, sometimes referred to as a *flexible manufacturing cell (FMC)*, consists of multiple machines, automated random movement of palletized parts to and from processing stations, and central computer control with sophisticated, command-driven software. The distinguishing characteristics of an FMS are: central computer control; the automated flow of raw material to the cell; complete machining of the part; part washing, drying, and inspection within the cell; and removal of the finished part.

An FMS is made up of hardware and software elements. Hardware elements, like CNC machine tools, part cleaning stations, coordinate measuring machines, and computers are visible and you can touch them. Software elements, like CNC

> **"Tech Talk"**
>
> FMS originated in London, England in the 1960s when David Williamson came up with a flexible machining system called "System 24" to operate unattended for 24 hours a day under computer control.

FIGURE 35–5
An in-line cell that progressively moves palletized parts from one machining station to another. (Courtesy of Cincinnati Milacron, Inc.)

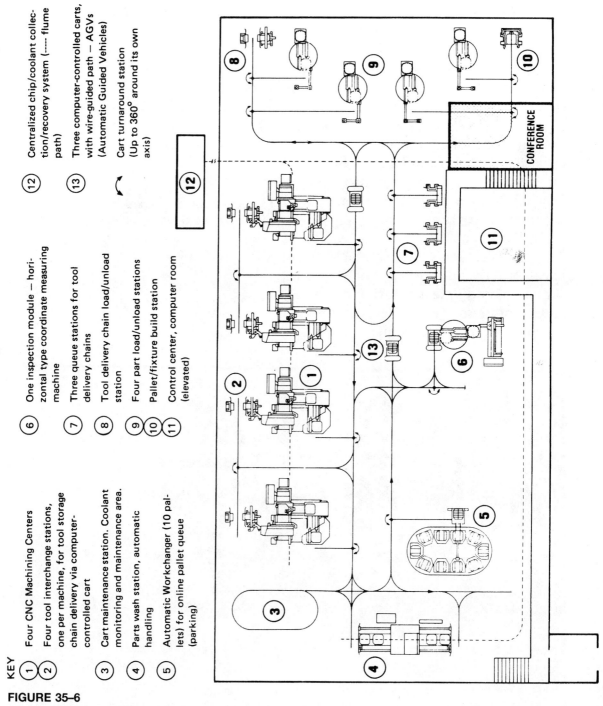

FIGURE 35–6
A typical FMS layout. (Courtesy of Cincinnati Milacron, Inc.)

KEY
① Four CNC Machining Centers
② Four tool interchange stations, one per machine, for tool storage chain delivery via computer-controlled cart
③ Cart maintenance station. Coolant monitoring and maintenance area.
④ Parts wash station, automatic handling
⑤ Automatic Workchanger (10 pallets) for online pallet queue (parking)
⑥ One inspection module — horizontal type coordinate measuring machine
⑦ Three queue stations for tool delivery chains
⑧ Tool delivery chain load/unload station
⑨ Four part load/unload stations
⑩ Pallet/fixture build station
⑪ Control center, computer room (elevated)
⑫ Centralized chip/coolant collection/recovery system (----- flume path)
⑬ Three computer-controlled carts, with wire-guided path — AGVs (Automatic Guided Vehicles)
↻ Cart turnaround station (Up to 360° around its own axis)

FIGURE 35-7
A full-scale, operational FMS. (Courtesy of Cincinnati Milacron, Inc.)

programs, tooling information, and work order files, are invisible and you cannot touch them. A typical FMS layout is shown in Figure 35-6.

A true FMS, as shown in Figures 35-7 and 35-8, can handle a variety of different parts, producing them one at a time in any order, as needed (very few FMSs meet this strict definition). To operate efficiently in this mode, an FMS must have several types of flexibility. It needs the flexibility to adapt to varying volume requirements and changing part mixes, to accept new parts, and to accommodate design and engineering changes. FMS also requires the flexibility to cope with unforeseen disturbances, such as machine downtime or last-minute schedule changes, and to grow and change with the times. These types of flexibility are made possible through computers and the FMS software that "drives" the entire system.

FIGURE 35–8
A true FMS can handle a variety of different parts producing them one at a time in any order, as needed. (Courtesy of Cincinnati Milacron, Inc.)

The following are advantages of an FMS:

- increased machine utilization (spindle running time)
- reduced work in process
- increased part throughput
- improved quality
- reduced inventory
- reduced personnel
- accurate scheduling
- reduced lead time

"Tech Talk"

The fundamental requirement with either a cell or FMS is to target the part family or families to be produced in order to scope out and determine the size, type, and layout of the cell or system.

Almost all low- and mid-volume manufacturers who face worldwide competition will need some form of flexible manufacturing to improve efficiency and effectiveness on the shop floor. Ultimately, they will want to apply the principles of flexible manufacturing throughout their operations.

RECALLING THE FACTS

1. In manufacturing terms, a _____ means arranging or organizing a cluster of manual or CNC machines into logical groups in order to process similar parts or a "family" of parts.

2. Today the term _____ _____ is much broader and, because of the advancements made in CNC and automation technologies, can mean some level of automated part and cutting tool loading, unloading, delivery, or exchange to the clustered machines.

3. _____ are moveable work tables that can be moved by hand or an automated process from machine to machine.

4. The _____ is traditonally a limited-tool and part storage automatic tool changer and is usually operated with one operator running one machine.

5. The _____, or _____ as it is sometimes called, consists of an automatic work changer with permanently assigned work pallets or a conveyor robot arm system mounted to the front of the machine.

6. The _____ cell is made up of multiple metal-cutting machine tools which can be of the same or a different type.

7. The _____, or _____, for short, consists of multiple machines, random movement of palletized parts to and from processing

stations, and central computer control with sophisticated command-driven software. List the advantages below.

- _____
- _____
- _____
- _____
- _____
- _____
- _____
- _____

UNIT 36

Automation Concepts and Capabilities

KEY TERMS

automation

automated storage and retrieval systems (ASRS)

conveyors

robots

automated guided vehicles (AGV)

AUTOMATION AND MODERN MANUFACTURING

The dictionary defines *automation* as "acting or operating in a manner essentially independent of external influence or control; self-moving." Just like an automatic transmission in a car that shifts itself, CNC machines, by their very nature, can be considered "automated." This is because, once programmed, they "act or operate independent of external influence or control" and are self-moving.

In Unit 35 you studied four levels of cells. These are ranked from the lowest level of automation (single stand-alone machine cell) to the highest level of automation (flexible manufacturing cell or system). As more capability is added to first control material flow, movement, and storage, and then perhaps tool, fixture, and pallet movement along with chip collection and removal, the cell takes on a higher and higher level of automation. The increasing automation then allows the cell or system to operate unattended or lightly attended.

Efforts to automate are driven by a number of forces and pressures acting on manufacturing industries today. These include, but are not limited to, the following:

- A declining percentage of young people choosing careers in manufacturing.
- Increased production pressure from foreign competition.
- Factory inefficiencies that create costly in-process inventory and tie up a manufacturer's money.

- Increased availability of automated systems for part loading and unloading.
- Significant advancements in electronics and microprocessor technology.
- The need for predictable and dependable productivity levels.

> **"Tech Talk"**
> Computer integrated manufacturing (CIM) is a systems approach toward manufacturing and business functions of an organization through the use of computers and commonly shared, easily accessible data for the purpose of improving company effectiveness.

Adding automation to a manufacturing plant utilizing CNC machines usually involves some of the following equipment: robots, automated guided vehicles (AGVs), automated storage and retrieval systems (ASRSs), and conveyors. You will now learn what these are, what they do, and how they help manufacturers automate their manufacturing processes.

ROBOTS

The industrial robot is one of the most important developments in the history of automation technology. Although the first robot arm was developed in the 1950s, the industrial robot, as we know it today, did not see wide-scale applications until the development of the microprocessor.

Robots, like machine tools, are available in a variety of types, styles, and sizes, as seen in Figures 36–1, 36–2, and 36–3. Generally, they are described as either nonservo or servo. Nonservo robots have no servo drives and have limited control over speeding up or slowing down movement. They are very simple, inexpensive devices with limited capability. Nonservo robots have two to four axes of movement, generally with one of the axes rotary, and they move one axis at a time. They operate either by compressed air or hydraulic actuation. Typically, they pick up an object and move it from point A to point B and are controlled by moving against a limited number of hard stops, trip points, or similar devices along each axis.

Servo-driven robots are determined by their electric or hydraulic servo drives. Their capability of being programmed makes them

AUTOMATION CONCEPTS AND CAPABILITIES • 355

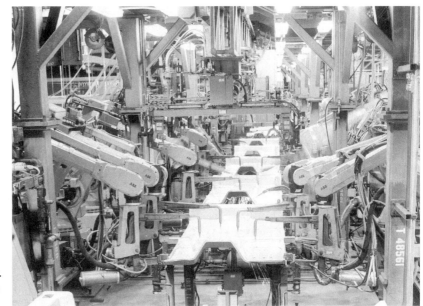

FIGURE 36–1
A modern industrial robot. (Courtesy of ABB Flexible Automation, Inc.)

FIGURE 36–2
Robots are available in a variety of types, styles, and sizes. (Photo: Courtesy of FANUC Robotics North America, Inc.)

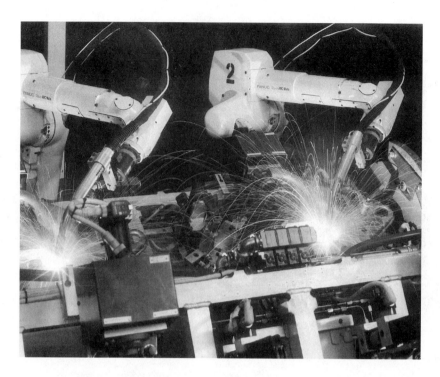

FIGURE 36–3
Another example of an industrial robot. (Photo: Courtesy of FANUC Robotics North America, Inc.)

FIGURE 36–4
Servo-driven robots are generally teach-and-lead-through type with the hand-held teach pendant or they may be programmed off-line. (Photo: Courtesy of FANUC Robotics North America, Inc.)

very flexible and open to a wide variety of applications. Servo-driven models generally are teach-and-lead-through type with the hand-held pendant (Figure 36–4), or they may be programmed off-line, much like CNC machines. They are the most commonly used robots because they are reprogrammable and can perform a variety of functions. Industrial robots are generally classified by:

1. arm configuration and reach
2. power sources and programmable speed
3. load capacity
4. application capabilities
5. control technique and intelligence

Just like a CNC machine, robots have axes of movement. The six primary axes of robot movement are shown in Figure 36–5. Also, methods of mounting can vary depending on the application. Although some robots are moveable and can be mounted on a cart that can traverse back and forth, most are fixed in a floor, wall, machine, or gantry mounting.

Benefits of using robots include:

- increased productivity
- improved product quality

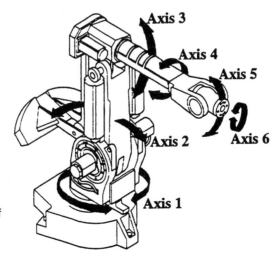

FIGURE 36–5
The six primary axes of robot movement. (Courtesy of ABB Flexible Automation, Inc.)

- application consistency
- increased operating hours
- reduced scrap and rework costs
- direct labor savings
- improved worker safety and job enrichment

Use of robots depends primarily on the application. Businesses typically do not purchase robots for the sake of having and using robots. Robots are generally purchased where applications are real, the environments are hostile, strenuous, repetitive, or dull, and the economic pressures to perform are high. Typical robot applications

FIGURE 36–6
Robots are purchased where applications are real, and the environments are hostile, strenuous, repetitive, or dull. (Courtesy of ABB Flexible Automation, Inc.)

FIGURE 36–7
A typical robot application. (Courtesy of ABB Flexible Automation, Inc.)

include: spot welding, spray painting, pick-and-place material handling, gluing, and others (Figures 36–6 and 36–7).

Robots can be interfaced with machine tools (Figure 36–8) for automated and progressive part loading and unloading. Conveyor systems can also be included to handle incoming and outgoing parts (Figure 36–9), along with interfacing robots with gauging systems for part checking and vision systems for optical recognition.

Other functions that robots are being used for in cells and systems include:

1. Loading and unloading parts on transporter devices or pallets in the staging area of an FMS.
2. Loading and unloading new or replacement tools in the tool-storage magazines of machine tools.
3. Deburring and cleaning completed parts.

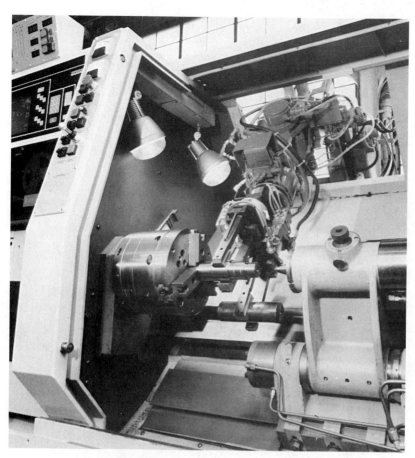

FIGURE 36-8
Robots can be interfaced with machine tools for automated part loading and unloading. (Courtesy of Cincinnati Milacron, Inc.)

FIGURE 36-9
Conveyor systems can be added to interface with machine tools and robots to further automate production. (Courtesy of Cincinnati Milacron, Inc.)

AUTOMATED GUIDED VEHICLES (AGVS)

AGVs are automated carts used to move work-in-process and finished goods inventory to and from machining or assembly stations. In addition, they can be used to move tools, fixtures, pallets, supplies, and other materials. In the 1970s, computers affected AGV technology as microprocessors began to be installed on AGVs. Vehicles began to be able to travel in different directions, moving material between assembly lines, shipping docks, and machines and, at the same time, provide material tracking.

AGVs come in a variety of types and sizes, as seen in Figures 36–10, 36–11, and 36–12 and can be used in applications and environments wherever material is moved. AGV general types include:

- towing
- pallet trucks

FIGURE 36–10
A modern AGV. (Courtesy of Cincinnati Milacron, Inc.)

FIGURE 36–11
AGVs can be used in applications and environments wherever material is moved. (Courtesy of FMC Corporation, Automated Material Handling Systems Division.)

FIGURE 36–12
A typical AGV application. (Courtesy of FMC Corporation, Automated Material Handling Systems Division.)

- unit load
- fork trucks
- assembly vehicles

AGVs typically can carry from 35 to 120,000 pounds at speeds ranging from 40 to 200 feet per minute. AGVs are becoming increasingly complex and sophisticated devices. Many are guided by onboard computers and some have vocabularies of up to 4,000 words and can be programmed in any language.

Many AGVs in use are battery-powered, wire-guided vehicles that follow energized wire embedded in the floor. Saw cuts are made in the floor about one-half inch deep based on a predetermined guide path route. Wire is laid in the saw cut and then epoxied over to form a smooth, unbroken surface for sweeping and maintenance. Guide path routes can conform to virtually any type of shop or warehouse layout. Carts move by electric motors that are powered by industrial-grade, lead-acid storage batteries mounted in the AGV. To some degree, wire-guided AGVs operate just like a model train set, only the track is invisible. Other means of guidance for AGVs range from teach-type and optical AGVs to those using laser scans, infrared light, and inertial guidance systems.

AGVs can be used in a variety of applications, from dirty manufacturing shops to clean-room assembly environments. They offer several advantages over conventional material-handling systems, including:

1. dispatching, tracking, and monitoring under real-time computer control,
2. increased control over material flow and movement,
3. reduced product damage and less material movement noise,
4. routing consistency but flexibility,
5. reliability in hazardous and special environments,
6. ability to interface other machines, robots, and conveyor systems,
7. high location and positioning accuracy, and
8. improved cost savings through reductions in floor space, work in process, and direct labor.

AUTOMATED STORAGE AND RETRIEVAL SYSTEMS (ASRSs)

Automated storage and retrieval systems, commonly referred to as ASRSs, are automated inventory-handling systems designed to replace manual and remote-control systems. Typically, they contain tall, vertical storage racks, narrow aisles, and stacker cranes and are coupled with some type of computer control, as seen in Figures 36–13 and 36–14. For the most part, ASRSs are strictly warehouse tools that track incoming material and components, store parts, tools, and fixtures, and retrieve them when necessary.

FIGURE 36–13
An automatic storage and retrieval system (ASRS). (Courtesy of FMC Corporation, Automated Material Handling Systems Division.)

AUTOMATION CONCEPTS AND CAPABILITIES • **365**

FIGURE 36–14
An ASRS is a warehouse tool that tracks incoming materials and components, stores parts, tools and fixtures, and retrieves them when necessary. (Photo courtesy of Jervis B. Webb Company.)

The goal of an automated storage and retrieval system is to deliver the right material to the right place at the right time. Material is held in storage and then issued to the point of use as close to the time of use as possible.

Automated storage and retrieval systems store standard-sized pallets of material, and have aisles that divide the storage racks. In each aisle is an arm or crane, sometimes known as a *stacker crane*. The crane picks up a load from an input station, stores it in a computer-assigned location, and delivers it to an output station, as seen in Figure 36–15. Stacker cranes are rated in terms of vertical and horizontal movement in feet per second. Cranes are capable of vertical and horizontal movement at the same time. Loads must be prepared

FIGURE 36–15
The automated crane of an ASRS will pick up a load from an input station, store in a computer-assigned location, and deliver to an output station. (Courtesy of FMC Corporation, Automated Material Handling Systems Division.)

for crane movement within size and weight limits, and they must be centered, as an off-center load can jam the crane.

The principal benefits of an ASRS are:

- improved inventory management and control
- reliable and immediate delivery
- space efficiency
- simplified and faster inventory response
- ability to operate in harsh environments
- reduced number of lost or misplaced parts, tools, and fixtures
- reduced labor costs
- accurate inventory and load location

CONVEYORS

Conveyors offer automated manufacturing users a variety of options from which to choose, depending on individual part characteristics and production requirements. They present a fixed path over which components travel to their destination. Conveyors are generally classified as either *overhead mounted* (Figure 36–16) or *floor mounted* (Figure 36–17). Overhead conveyors may be either of the monorail

FIGURE 36–16
An overhead mounted conveyor. (Photo courtesy of Jervis B. Webb Company.)

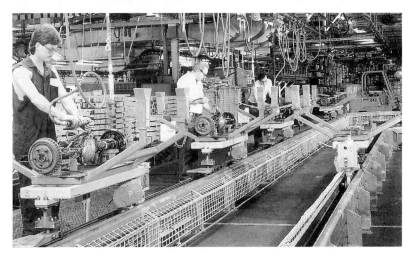

FIGURE 36–17
A floor mounted conveyor. (Photo courtesy of Jervis B. Webb Company.)

FIGURE 36–18
Chain and roller conveyors are used for a wide variety of applications and accommodate varying types of loads. (Photo courtesy of Jervis B. Webb Company.)

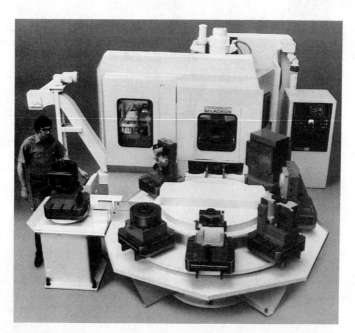

FIGURE 36–19
A carrousel conveyor. (Courtesy of Cincinnati Milacron, Inc.)

power-and-free type or overhead chain type. Both power-and-free and chain-driven conveyors can handle medium to large part types such as automobile frames and bodies.

Floor-mounted conveyors are classified as *chain, roller,* or *belt-driven*. Chain and roller conveyors are very practical and can accommodate varying types of loads (Figure 36–18). They are sometimes used with free-floating pallets in dedicated manufacturing systems or in group technology cells where workpieces are passed on pallets from one machine to another. Carrousel conveyors (Figure 36–19) are sometimes belt-driven and typically handle small parts. They are often used in loop applications in manufacturing cells to present parts to a robot for transferring to a machine tool, as seen in Figure 36–20.

FIGURE 36–20
Carrousel conveyors are belt-driven and designed to handle small parts. (Courtesy of ABB Flexible Automation, Inc.)

RECALLING THE FACTS

1. The dictionary defines _____ as "acting or operating in a manner essentially independent of external influence or control; self-moving."
2. The industrial _____, as we know it today, did not see wide-scale applications until the development of the _____.
3. _____ robots have limited control over speeding up or slowing down movement; _____-driven robots are determined by their electric or hydraulic _____ motors.
4. Industrial _____ are determined by their:
   a. configuration and reach,
   b. sources and programmable speed,
   c. _____,
   d. _____ capabilities, and
   e. control technique and _____.
5. Robots have _____ primary axes of motion.
6. Benefits of using robots are:
   - _____
   - _____
   - _____
   - increased operating hours
   - _____
   - _____
   - _____
7. Typical robot applications are _____, _____, _____, and gluing.
8. _____ are automated carts used to move work-in-process and finished goods inventory to and from machining or assembly stations.
9. The majority of _____ in use are battery-powered and follow an energized _____ embedded in the floor.
10. The advantages of _____ over conventional material-handling equipment include:

a. dispatching, _____, and monitoring under real-time computer control,
b. increased _____ over material flow and _____,
c. reduced product _____ and less material movement _____,
d. routing _____ but flexibility,
e. reliability in hazardous and special _____,
f. ability to _____ other machines, robots, and conveyor systems,
g. high location and positioning _____, and
h. improved cost _____ through reductions in _____ space, work in process, and _____ labor.

11. _____ consist of tall, vertical storage racks, narrow aisles, and stacker cranes and are coupled with some type of computer control.

12. The goal of an _____ is to deliver the right _____ to the right _____ at the right _____.

13. The principal benefits of automated storage and retrieval systems are:

- _____
- _____
- _____
- simplied and faster inventory response
- _____
- _____
- accurate inventory and load location

14. _____ present a fixed path over which components travel to their destination. _____ are classified as either _____ mounted or _____ mounted.

15. Floor-mounted conveyors are classified as either _____, _____, or _____-driven.

SECTION 6
REVIEW

KEY POINTS

- CNC hardware has advanced considerably, primarily because of the development of the 32-bit microprocessor, which has improved the processing speed up to 10 times faster than the older 16-bit CNC control.
- Shop floor programming, or SFP, is an automated part-programming system embedded in a CNC that enables programming input to be in conversational English.
- Many advanced CNC applications are in use today beyond the traditional metal-cutting, laser drilling, and forming. These include CNC machines for composite lay-up, waterjet, and abrasive waterjet machining.
- APT (automatic programmed tool) was developed in the early 1960s and is still the most common and widely used processor language today for N/C applications.

I remember that quote early in the text: "man is a user of tools; those who recognize the tools of tromorrow and learn how to use them today assure themselves of a place in tomorrow's prosperity." Now I think I really understand.

- APT is an English-like language that has its own vocabulary and must be learned like any other language.
- APT part programming involves three major elements: (1) defining and naming the part's geometric points and elements, (2) specifying the cutting tool and tool motion, and (3) specifying required machine tool conditions.
- The output from APT is called centerline *cutter location*, or CL data. CL data must be converted into machine language code that a CNC unit can understand by a postprocessor.
- CAD/CAM systems allow for detail parts and assemblies to be designed with part geometry being stored electronically in a central database.
- The vast majority of CAD usage is 2-D (just like part prints). CAD systems are primarily used to create lines, surfaces, solids, intersections, and curved surfaces.
- Graphics-based systems are state-of-the-art today and were introduced to simplify the programming process, move visualization from the programmer's mind to the computer screen, and provide immediate feedback to the programmer.
- Graphics-based systems permit creating a part by drawing lines, circles, arcs, and planes and provide the programmer with the ability to visually and graphically follow the tool path by obtaining a three-dimensional view of the part and cutter clearance planes.
- In manufacturing terms, a *cell* means arranging or organizing a cluster of manual or CNC machines into logical groups in order to process similar parts or a "family" of parts.
- Manufacturing cells can be divided into four general categories: (1) traditional stand-alone CNC machine cells, (2) single CNC machine cells, or minicells, (3) multi-machine cells, (4) flexible and manufacturing systems (FMSs).
- Efforts to automate are driven by a number of forces and pressures acting on manufacturing industries today. These include, but are not limited to, the following: young people not choosing careers in manufacturing, more foreign competition, factory inefficiencies, more available and affordable automation capabilities, advancements in electronics and micro-

processor technology, and the need for predictable and dependable productivity levels.
- Some automation concepts and capabilities helping manufacturers to automate and improve productivity and reduce waste are: robots, automated guided vehicles (AGVs), automated storage and retrieval systems (ASRSs), and conveyors.

REVIEW QUESTIONS AND EXERCISES

1. Development of the _____ microprocessor has improved the processing speed of CNCs, making them 10 times faster than the 16-bit control.
2. List the four parts of a CNC that work together to provide improved CNC machine tool performance.
3. _____ is an automated part-programming system embedded in a CNC where programming input can be in conversational English.
4. Language-based processors cannot produce graphics displays. (True/False)
5. APT part programming involves specifying the conditions required at the machine tool, such as spindle speeds, specifying the cutting tool and tool motion, and
   a. identifying all of the points to be machined.
   b. figuring out the coordinates of the intersections.
   c. defining and naming the part's geometric points and surfaces.
   d. figuring out the motion statements.
   e. all of the above
6. The output from APT is called _____.
7. A _____ is a set of computer instructions which transforms tool centerline data into machine motion commands.
8. CAD/CAM systems allow for detail parts and assemblies to be designed with part geometry being stored electronically in a central _____.
9. Language-based computer part programming is simpler and easier to learn than graphics-based part programming. (True/False)

10. A manufacturing _____ can be arranged by part type, machine function, color, or however a manufacturer wants it to be organized.
11. Of the four types of manufacturing cells, a flexible manufacturing system contains the highest level of automation and is the most costly. (True/False)
12. An FMS is made up of hardware and _____ elements.
13. List four reasons why manufacturers are being driven to automate.
14. Name four types of automation equipment that manufacturing companies are adding in order to automate their facilities.
15. A robot has _____ primary axes of movement.
16. List and explain five new words you learned in this chapter.
17. Using full sentences and in your own words, write a paragraph about what forces and pressures to automate are acting on companies today.

SHOP ACTIVITIES

1. Identify two opportunities in your shop where automation could make an improvement. Make a basic writeup or proposal of the concept along with basic diagrams relative to shop floor layout changes, etc. Get help from engineers, manufacturing engineers, or others who are knowledgeable in this area. If you are in a school machine tool lab you can focus on a machine or group of machines in the school shop that could be automated if you were going to run production parts. Use your imagination.
2. Talk with your supervisor or instructor about having an automation vendor come in as a guest speaker. An automation vendor could be someone from a machine tool company (cells and systems), a conveyor company, ASRS, AGV, tool holder/delivery manufacturer, etc. If one is not available, write away for information and have a group review and discussion about the material.

APPENDIX A

EIA AND AIA NATIONAL CODES

Preparatory Functions

G word	Explanation
G00	Denotes a rapid traverse rate for point-to-point positioning.
G01	Describes linear interpolation blocks; reserved for contouring.
G02, G03	Used with circular interpolation.
G04	Sets a calculated time delay during which there is no machine motion (dwell).
G05, G07	Unassigned by the EIA. May be used at the discretion of the machine tool or system builder. Could also be standardized at a future date.
G06	Used with parabolic interpolation.
G08	Acceleration code. Causes the machine, assuming it is capable, to accelerate at a smooth exponential rate.
G09	Deceleration code. Causes the machine, assuming it is capable, to decelerate at a smooth exponential rate.
G10–G12	Normally unassigned for CNC systems. Used with some hard-wired systems to express blocks of abnormal dimensions.
G13–G16	Direct the control system to operate on a particular set of axes.
G17–G19	Identify or select a coordinate plane for such functions as circular interpolation or cutter compensation.
G20–G32	Unassigned according to EIA standards. May be assigned by the control system or machine tool builder.
G33–G35	Selected for machines equipped with thread-cutting capabilities (generally referring to lathes). G33 is used when a constant lead is sought, G34 is used when a constantly increasing lead is required, and G35 is used to designate a constantly decreasing lead.
G36–G39	Unassigned.
G40	Terminates any cutter compensation.
G41	Activates cutter compensation in which the cutter is on the left side of the work surface (relative to the direction of the cutter motion).
G42	Activates cutter compensation in which the cutter is on the right side of the work surface.

G43, G44	Used with cutter offset to adjust for the difference between the actual and programmed cutter radii or diameters. G43 refers to an inside corner, and G44 refers to an outside corner.
G45–G49	Unassigned.
G50–G59	Reserved for adaptive control.
G60–G69	Unassigned.
G70	Selects inch programming.
G71	Selects metric programming.
G72	Selects three-dimensional CW circular interpolation.
G73	Selects three-dimensional CCW circular interpolation.
G74	Cancels multiquadrant circular interpolation.
G75	Activates multiquadrant circular interpolation.
G76–G79	Unassigned.
G80	Cancel cycle.
G81	Activates drill, or spotdrill, cycle.
G82	Activates drill with a dwell.
G83	Activates intermittent, or deep-hole, drilling.
G84	Activates tapping cycle.
G85–G89	Activate boring cycles.
G90	Selects absolute input. Input data is to be in absolute dimensional form.
G91	Selects incremental input. Input data is to be in incremental form.
G92	Preloads registers to desired values (for example, preloads axis position registers).
G93	Sets inverse time feed rate.
G94	Sets inches (or millimeters) per minute feed rate.
G95	Sets inches (or millimeters) per revolution feed rate.
G97	Sets spindle speed in revolutions per minute.
G98, G99	Unassigned.

Miscellaneous Functions

M word	Explanation
M00	Program stop. Operator must cycle start in order to continue with the remainder of the program.
M01	Optional stop. Acted upon only when the operator has previously signaled for this command by pushing a button. When the control system senses the M01 code, machine will automatically stop.
M02	End of program. Stops the machine after completion of all commands in the block. May include rewinding of tape.
M03	Starts spindle rotation in a clockwise direction.
M04	Starts spindle rotation in a counterclockwise direction.
M05	Spindle stop.

M06	Executes the change of a tool (or tools) manually or automatically.
M07	Turns coolant on (flood).
M08	Turns coolant on (mist).
M09	Turns coolant off.
M10	Activates automatic clamping of the machine slides, workpiece, fixture, spindle, etc.
M11	Deactivates automatic clamping.
M12	Inhibiting code used to synchronize multiple sets of axes, such as a four-axis lathe that has two independently operated heads or slides.
M13	Combines simultaneous clockwise spindle motion and coolant on.
M14	Combines simultaneous counterclockwise spindle motion and coolant on.
M15	Sets rapid traverse or feed motion in the + direction.
M16	Sets rapid traverse or feed motion in the − direction.
M17, M18	Unassigned.
M19	Oriented spindle stop. Stops spindle at a predetermined angular position.
M20–M29	Unassigned.
M30	End of data. Used to reset control and/or machine.
M31	Interlock bypass. Temporarily circumvents a normally provided interlock.
M32–M39	Unassigned.
M40–M46	Signals gear changes if required at the machine; otherwise, unassigned.
M47	Continues program execution from the start of the program, unless inhibited by an interlock signal.
M48	Cancels M49.
M49	Deactivates a manual spindle or feed override and returns to the programmed value.
M50–M57	Unassigned.
M58	Cancels M59.
M59	Holds the rpm constant at its value.
M60–M99	Unassigned.

Other Address Characters

Address character	Explanation
A	Angular dimension about the X-axis.
B	Angular dimension about the Y-axis.
C	Angular dimension about the Z-axis.
D	Can be used for an angular dimension around a special axis, for a third feed function, or for tool offset.
E	Used for angular dimension around a special axis or for a second feed function.
H	Fixture offset.
I, J, K	Centerpoint coordinates for circular interpolation.

L	Not used.
O	Used on some N/C controls in place of the customary sequence number word address N.
P	Third rapid traverse code—tertiary motion dimension parallel to the X-axis.
Q	Second rapid traverse code—tertiary motion dimension parallel to the Y-axis.
R	First rapid traverse code—tertiary motion dimension parallel to the Z-axis (or to the radius) for constant surface speed calculation.
U	Secondary motion dimension parallel to the X-axis.
V	Secondary motion dimension parallel to the Y-axis.
W	Secondary motion dimension parallel to the Z-axis.

APPENDIX B

USEFUL FORMULAS AND TABLES

Determining rpm

$$\text{rpm} = \frac{3.82 \times \text{Cutspeed}}{\text{Diameter}}$$

Switching between feed values in inches per minute (ipm) and inches per revolution (ipr)

$$\text{Feed (ipm)} = (\text{rpm}) \times \text{Feed (ipr)}$$

$$\text{Feed (ipr)} = \frac{\text{Feed (ipm)}}{\text{rpm}}$$

Useful calculations

- To find the circumference of a circle, multiply the diameter by 3.1416.
- To find the diameter of a circle, multiply the circumference by 0.31831.
- To find the area of a circle, multiply the square of the diameter by 0.7854.
- To obtain the circumference, multiply the radius of a circle by 6.283185.
- To obtain the area of a circle, multiply the square of the circumference of a circle by 0.07958.
- To find the area of a circle, multiply half the circumference of a circle by half its diameter.
- To obtain the radius of a circle, multiply the circumference of a circle by 0.159155.
- To find the radius of a circle, multiply the square root of the area of a circle by 0.56419.
- To find the diameter of a circle, multiply the square root of the area of a circle by 1.12838.
- To find the area of the surface of a ball (sphere), multiply the square of the diameter by 3.1416.
- To find the volume of a ball (sphere), multiply the cube of the diameter by 0.5236.

Trigonometry

$$c^2 = a^2 + b^2$$

$$c = \sqrt{a^2 + b^2} \quad a = \sqrt{c^2 - b^2} \quad b = \sqrt{c^2 - a^2}$$

where c = hypotenuse length of a right triangle
 a = length of one of the short sides of a right triangle
 b = length of the other short side

$$\text{Sine} = \frac{\text{Side opposite}}{\text{Hypotenuse}} \quad \text{Cosecant} = \frac{\text{Hypotenuse}}{\text{Side opposite}}$$

$$\text{Cosine} = \frac{\text{Side adjacent}}{\text{Hypotenuse}} \quad \text{Secant} = \frac{\text{Hypotenuse}}{\text{Side adjacent}}$$

$$\text{Tangent} = \frac{\text{Side opposite}}{\text{Side adjacent}} \quad \text{Cotangent} = \frac{\text{Side adjacent}}{\text{Side opposite}}$$

CUTTING SPEEDS
(Feet Per Minute)

MATERIAL	DRILL HSS	DRILL CARBIDE	REAM HSS	REAM CARBIDE	TAP HSS	COBORE HSS	COBORE CARBIDE	BORE HSS	BORE CARBIDE	MILLING HIGH SPEED Rgh.	MILLING HIGH SPEED Fin.	MILLING CARBIDE Rgh.	MILLING CARBIDE Fin.
Aluminum	200	350	175	300	90	180	300	300	600	240	300	500	1000
Brass — Soft	145	350	120	250	100	150	300	150	450	150	200	400	600
Brass — Hard	125	225	100	200	75	110	200	120	350	135	180	350	500
Bronze — Common	140	250	125	200	90	130	200	150	400	145	190	360	550
Bronze — High Tensile	60	200	50	175	40	55	180	85	300	70	90	200	280
Cast Iron — Soft 170 BHN	90	180	60	200	40	85	160	80	280	90	110	250	350
Cast Iron — Medium 220 BHN	60	140	45	125	30	55	130	55	255	70	90	200	300
Cast Iron — Hard 300 BHN	40	120	30	60	20	35	100	45	215	50	60	175	250
Cast Iron — Malleable	85	140	45	100	40	75	180	90	250	100	120	260	370
Cast Steel	60	120	50	100	40	60	180	70	200	50	80	225	380
Copper	75	250	50	125	40	70	200	95	350	90	150	220	400
Magnesium	250	500	180	450	150	200	450	400	1000	300	400	600	1000
Monel	50	100	35	90	20	45	90	50	110	60	80	180	240
Steel — Mild .2 to .3 Carbon	95	—	50	250	40	85	170	80	280	90	130	300	450
Steel — Medium .4 to .5 Carbon	75	—	45	200	35	60	120	80	220	70	85	210	400
Steel — Tool up to 1.2 Carbon	40	80	30	70	20	40	80	45	190	50	80	175	350
Steel — Forging	45	90	35	80	25	40	80	50	200	60	80	200	300
Steel — Alloy 300 BHN	60	120	40	115	35	60	120	70	250	60	80	250	350
Steel — Alloy 400 BHN	45	90	30	65	25	40	80	40	165	30	40	160	250
Steel — High Tensile to 40 Rc	35	70	30	60	20	30	60	40	150	40	50	120	150
Steel — High Tensile to 45 Rc	30	60	20	50	15	20	40	30	100	35	45	110	140
Steel — Stainless — Free Machining	55	110	35	100	25	50	100	50	150	40	60	200	400
Steel — Stainless — Work Hardening	30	60	20	50	15	30	60	40	90	30	50	180	300
Titanium — Commercially Pure	55	110	45	100	30	50	100	60	120	60	75	200	280
Zinc Die Casting	150	300	125	225	80	150	250	180	350	200	300	250	450

CONVERSION CHART
CUTTING SPEEDS TO RPM

	1/8	3/16	1/4	5/16	3/8	7/16	1/2	9/16	5/8	11/16	3/4	13/16	7/8	15/16	1	1-1/8	1-1/4	1-3/8	1-1/2	1-5/8	1-3/4	2	2-1/4	2-1/2	2-3/4	3	3-1/2	4	4-1/2	5	5-1/2	6	6-1/2	7	7-1/2	8	
10	306	204	153	122	102	87	76	68	61	56	51	47	44	41	38	34	31	28	25	24	22	19	17	15													10
20	611	407	306	244	204	175	153	136	122	111	102	94	87	82	76	68	61	56	51	47	44	38	34	31	28	25	22	19	17	15							20
30	917	611	458	368	306	262	229	204	183	167	153	141	131	122	115	102	92	83	76	71	65	57	51	46	42	38	33	29	25	23	21	19	18	16	15		30
40	1222	815	611	489	408	349	306	272	245	222	204	188	175	163	153	136	122	111	102	94	87	76	68	61	56	51	44	38	34	31	28	25	24	22	20	19	40
50	1528	1020	764	611	509	437	382	339	306	278	255	235	218	204	191	170	153	140	127	118	109	95	85	76	69	64	55	48	42	38	35	32	29	27	25	24	50
60	1834	1222	917	733	611	524	458	407	367	333	306	282	262	245	229	204	183	167	153	141	131	115	102	92	83	76	65	57	51	46	42	38	35	33	31	29	60
70	2140	1426	1070	856	713	611	535	475	428	389	357	329	306	285	267	238	214	194	178	165	153	134	119	107	97	89	76	67	59	53	49	45	41	38	36	33	70
80	2445	1630	1222	978	813	700	611	543	489	444	408	376	350	326	306	272	244	222	204	188	175	153	136	122	111	102	87	76	68	61	56	51	47	44	41	38	80
90	2750	1833	1375	1100	917	786	688	611	550	500	458	423	393	367	344	306	275	250	229	212	196	172	153	138	125	115	98	86	76	69	63	57	53	49	46	43	90
100		2037	1528	1222	1020	873	764	679	611	556	509	470	436	408	382	340	306	278	255	235	218	191	170	153	139	127	109	96	85	76	70	64	59	55	51	48	100
120		2445	1834	1467	1222	1048	917	815	733	667	611	564	524	489	458	407	367	333	306	282	262	229	204	183	167	153	131	115	102	92	83	76	71	65	61	57	120
140		2852	2140	1711	1426	1222	1070	950	856	778	713	658	611	571	535	475	428	390	356	329	306	267	238	214	194	178	153	134	119	107	97	89	82	76	71	67	140
160			2445	1956	1630	1397	1222	1086	978	889	815	752	698	652	611	543	490	444	407	376	350	306	272	244	222	204	175	153	136	122	111	102	94	87	81	76	160
180			2750	2200	1834	1572	1375	1222	1100	1000	917	846	786	734	688	611	550	500	458	423	393	344	306	275	250	229	196	172	153	138	125	115	106	98	92	86	180
200				2445	2037	1747	1528	1358	1222	1111	1020	940	874	815	764	680	611	556	510	470	437	382	340	306	278	255	218	191	170	153	139	127	117	109	102	96	200
220				2690	2240	1920	1681	1494	1345	1222	1121	1034	960	897	840	747	672	611	560	517	480	420	374	336	306	280	240	210	187	168	153	140	129	120	112	105	220
240				2934	2445	2096	1833	1630	1467	1333	1222	1128	1048	978	917	815	733	667	611	564	524	458	407	367	333	306	262	229	204	183	167	153	141	131	122	115	240
250					2547	2183	1910	1697	1528	1389	1274	1175	1091	1020	955	850	764	694	637	588	546	477	424	382	347	318	273	240	212	191	174	159	147	136	127	119	250
260					2650	2270	1886	1765	1590	1444	1325	1221	1135	1060	993	883	795	722	662	611	568	497	441	397	361	331	284	221	200	200	181	166	153	142	132	124	260
280					2850	2445	2140	1900	1712	1556	1426	1316	1222	1140	1070	950	856	778	713	658	611	535	475	428	390	357	306	267	238	214	194	178	165	153	143	134	280
300						2620	2292	2037	1834	1667	1528	1410	1310	1222	1146	1020	917	833	764	705	655	573	510	458	417	382	327	287	255	229	208	191	176	164	153	143	300
320						2795	2445	2172	1956	1778	1630	1504	1397	1304	1222	1086	978	889	815	752	698	611	543	490	444	407	350	306	272	244	222	204	188	175	163	153	320
340						2970	2597	2308	2078	1889	1732	1600	1484	1385	1300	1155	1040	944	866	800	742	650	577	520	472	433	372	325	290	260	236	216	200	186	173	162	340
350							2674	2375	2140	1944	1783	1646	1528	1426	1337	1190	1070	972	891	823	764	668	594	535	486	446	381	334	297	267	243	223	206	191	178	167	350
360							2750	2444	2200	2000	1834	1693	1570	1467	1375	1222	1100	1000	917	846	786	688	611	550	500	458	393	344	306	275	250	230	212	196	183	172	360
380							2900	2580	2323	2111	1936	1787	1660	1550	1450	1290	1160	1055	968	893	830	725	645	580	528	484	415	363	323	290	264	242	223	207	194	181	380
400								2715	2445	2222	2038	1881	1746	1630	1530	1360	1222	1111	1020	940	873	764	680	611	556	510	437	382	340	305	278	255	235	218	204	191	400
450									2750	2500	2292	2116	1964	1834	1720	1530	1375	1250	1146	1060	982	860	764	688	625	573	491	430	382	344	313	286	264	246	230	215	450
500										2778	2550	2350	2180	2037	1910	1700	1530	1390	1273	1175	1090	955	850	764	694	637	546	478	425	382	347	318	294	273	255	239	500
550											2800	2586	2400	2240	2100	1870	1680	1530	1400	1293	1200	1050	934	840	764	700	600	525	467	420	382	350	325	300	280	263	550
600												2821	2620	2445	2290	2040	1834	1667	1528	1410	1310	1145	1020	917	833	764	655	573	510	458	417	382	353	327	306	287	600
	1/8	3/16	1/4	5/16	3/8	7/16	1/2	9/16	5/8	11/16	3/4	13/16	7/8	15/16	1	1-1/8	1-1/4	1-3/8	1-1/2	1-5/8	1-3/4	2	2-1/4	2-1/2	2-3/4	3	3-1/2	4	4-1/2	5	5-1/2	6	6-1/2	7	7-1/2	8	

TOOL DIAMETERS

TAPPING FEED RATES

THREADS PER INCH-TO-LEAD

TPI	LEAD	TPI	LEAD	TPI	LEAD
3	.3333	11-1/2	.0870	32	.0313
3-1/2	.2587	12	.0833	36	.0278
4	.2500	13	.0769	40	.0250
5	.2000	14	.0714	44	.0227
6	.1667	16	.0625	48	.0208
7	.1430	18	.0556	56	.0179
8	.1250	20	.0500	64	.0156
9	.1111	24	.0417	72	.0139
10	.1000	27	.0370	80	.0125
11	.0909	28	.0357		

Program feed rate = Lead of tap (inches) × rpm

CONVERSION CHART
(Based on 25.4 mm = 1")
Inches into Millimeters

Inches		M/M	Inches		M/M	Inches	M/M	Inches	M/M	Inches	M/M
1/64	.0156	0.3969	49/64	.7656	19.4469	34	863.600	82	2082.80	130	3302.00
1/32	.0313	0.7937	25/32	.7813	19.8437	35	889.000	83	2108.20	131	3327.40
3/64	.0469	1.1906	51/64	.7969	20.2406	36	914.400	84	2133.60	132	3352.80
1/16	.0625	1.5875	13/16	.8125	20.6375	37	939.800	85	2159.00	133	3378.20
5/64	.0781	1.9844	53/64	.8281	21.0344	38	965.200	86	2184.40	134	3403.60
3/32	.0938	2.3812	27/32	.8438	21.4312	39	990.600	87	2209.80	135	3429.00
7/64	.1094	2.7781	55/64	.8594	21.8281	40	1016.00	88	2235.20	136	3454.40
1/8	.1250	3.1750	7/8	.8750	22.2250	41	1041.40	89	2260.60	137	3479.80
9/64	.1406	3.5719	57/64	.8906	22.6219	42	1066.80	90	2286.00	138	3505.20
5/32	.1563	3.9687	29/32	.9063	23.0187	43	1092.20	91	2311.40	139	3530.60
11/64	.1719	4.3656	59/64	.9219	23.4156	44	1117.60	92	2336.80	140	3556.00
3/16	.1875	4.7625	15/16	.9375	23.8125	45	1143.00	93	2362.20	141	3581.40
13/64	.2031	5.1594	61/64	.9531	24.2094	46	1168.40	94	2387.60	142	3606.80
7/32	.2188	5.5562	31/32	.9688	24.6062	47	1193.80	95	2413.00	143	3632.20
15/64	.2344	5.9531	63/64	.9844	25.0031	48	1219.20	96	2438.40	144	3657.60
1/4	.2500	6.3500	1		25.4000	49	1244.60	97	2463.80	145	3683.00
17/64	.2656	6.7469	2		50.800	50	1270.00	98	2489.20	146	3708.40
9/32	.2813	7.1437	3		76.200	51	1295.40	99	2514.60	147	3733.80
19/64	.2969	7.5406	4		101.600	52	1320.80	100	2540.00	148	3759.20
5/16	.3125	7.9375	5		127.000	53	1346.20	101	2565.40	149	3784.60
21/64	.3281	8.3344	6		152.400	54	1371.60	102	2590.80	150	3810.00
11/32	.3438	8.7312	7		177.800	55	1397.00	103	2616.20	151	3835.40
23/64	.3594	9.1281	8		203.200	56	1422.00	104	2641.60	152	3860.80
3/8	.3750	9.5250	9		228.600	57	1447.80	105	2667.00	153	3886.20
25/64	.3906	9.9219	10		254.000	58	1473.20	106	2692.40	154	3911.60
13/32	.4063	10.3187	11		279.400	59	1498.60	107	2717.80	155	3937.00
27/64	.4219	10.7156	12		304.800	60	1524.00	108	2743.20	156	3962.40
7/16	.4375	11.1125	13		330.200	61	1549.40	109	2768.60	157	3987.80
29/64	.4531	11.5094	14		355.600	62	1574.80	110	2794.00	158	4013.20
15/32	.4688	11.9062	15		381.000	63	1600.20	111	2819.40	159	4038.60
31/64	.4844	12.3031	16		406.400	64	1625.60	112	2844.80	160	4064.00
1/2	.5000	12.7000	17		431.800	65	1651.00	113	2870.20	161	4089.40
33/64	.5156	13.0969	18		457.200	66	1676.40	114	2895.60	162	4114.80
17/32	.5313	13.4937	19		482.600	67	1701.80	115	2921.00	163	4140.20
35/64	.5469	13.8906	20		508.000	68	1727.20	116	2946.40	164	4165.60
9/16	.5625	14.2875	21		533.400	69	1752.60	117	2971.80	165	4191.00
37/64	.5781	14.6844	22		558.800	70	1778.00	118	2997.20	166	4216.40
19/32	.5938	15.0812	23		584.200	71	1803.40	119	3022.60	167	4241.80
39/64	.6094	15.4781	24		609.600	72	1828.80	120	3048.00	168	4267.20
5/8	.6250	15.8750	25		635.000	73	1854.20	121	3073.40	169	4292.60
41/64	.6406	16.2719	26		660.400	74	1879.60	122	3098.80	170	4318.00
21/32	.6563	16.6687	27		685.800	75	1905.00	123	3124.20	171	4343.40
43/64	.6719	17.0656	28		711.200	76	1930.40	124	3149.60	172	4368.80
11/16	.6875	17.4625	29		736.600	77	1955.80	125	3175.00	173	4394.20
45/64	.7031	17.8594	30		762.000	78	1981.20	126	3200.40	174	4419.60
23/32	.7188	18.2562	31		787.400	79	2006.60	127	3225.80	175	4445.00
47/64	.7344	18.6531	32		812.800	80	2032.00	128	3251.20		
3/4	.7500	19.0500	33		838.200	81	2057.40	129	3276.60		

0.001" = .0254 mm   0.001 mm = 0.0004"

GLOSSARY

A-AXIS. (or α-axis) The axis of circular motion of a machine tool member or slide about the X-axis. Values along the A-axis are degrees of rotation about the X-axis.

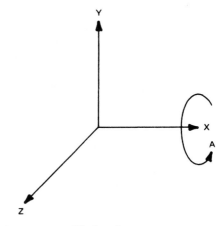

ABSOLUTE ACCURACY. Accuracy as measured from a specified reference.

ABSOLUTE READOUT. A display of the true slide position as derived from the position commands within the control system.

ABSOLUTE SYSTEM. A numerical control system in which all positional dimensions, both input and feedback, are given with respect to a common datum point. The alternative is the incremental system.

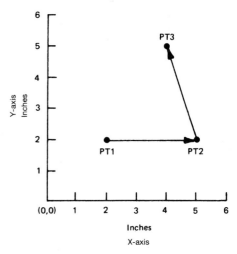

In an absolute system, all points are relative to (0,0), and the absolute coordinates for each of the required points are programmed with respect to (0,0).

ACCANDEC. (acceleration and deceleration) Acceleration and deceleration in feed rate. It provides smooth starts and stops when operating in N/C and when changing from one feed rate value to another.

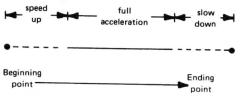

Most modern numerical control systems have automatic acceleration and deceleration.

ACCURACY. 1. A measure of the difference between the actual position of the machine slide and the position demanded. 2. Conformity of an indicated value to a true value (that is, an actual or an accepted standard value). The accuracy of a control system is expressed as the deviation (the difference between the ultimately controlled variable and its ideal value), usually in the steady state or at sampled instants.

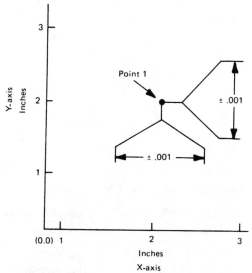

The position of point 1 in this example is X = 2 and Y = 2. If the machine accuracy is specified as ± .001, the X-axis movement could be between X = 1.999 and X = 2.001. The Y-axis movement could be between Y = 1.999 and Y = 2.001.

C1 = circle/center, PT1, radius, 2.5

Similar to the APT language except it does not possess the advanced contouring capabilities of APT.

AD-APT. An Air Force adaptation of APT program language, with limited vocabulary. It can be used for N/C programming on some small- to medium-size U.S. computers.

ADAPTIVE CONTROL. A technique that automatically adjusts feeds and/or speeds to an optimum by sensing cutting conditions and acting upon them.

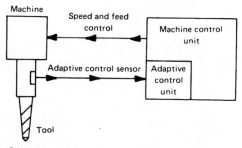

Sensors may measure variable factors, e.g. vibration, heat, torque, and deflection. Cutting speeds and feeds may be increased or decreased depending on conditions sensed.

ADDRESS. 1. A symbol indicating the significance of the information immediately following. 2. A means of identifying information or a location in a control system. 3. A number that identifies one location in memory.

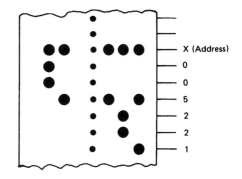

ALGORITHM. 1. A computational procedure for solving a problem. When properly applied, an algorithm always produces a solution to a problem.

ALPHANUMERIC CODING. A system in which the characters are letters A through Z and numerals 0 through 9.

APT and AD-APT statements use alphanumeric coding, e.g. GOFWD, CT12/PAST, 2, INTOF, L13

ANALOG. 1. Pertaining to a system that uses electrical voltage magnitudes or ratios to represent physical axis positions. 2. Pertaining to information that can have continuously variable values.

ANALYST. A person skilled in the definition and development of techniques to solve problems.

APT. (Automatic Programmed Tool) A universal computer-assisted program system for multiaxis contouring programming. APT III provides for five axes of machine tool motion.

Typical APT geometry definition statement:
  C1 = CIRCLE/XLARGE, L12, XLARGE, L13,
  RADIUS, 3.5

Typical APT tool motion statement:
  TLRGT, GORGT/AL3, PAST, AL12

ARC CLOCKWISE. An arc, generated by the coordinated motion of two axes, in which curvature of the tool path with respect to the workpiece is clockwise when the plane of motion is viewed from the positive direction of the perpendicular axis.

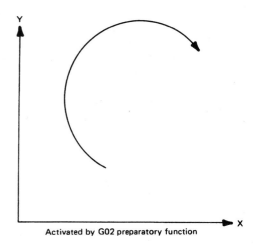

Activated by G02 preparatory function

ARC COUNTERCLOCKWISE. An arc, generated by the coordinated motion of two axes, in which curvature of the tool path with respect to the workpiece is counterclockwise when the plane of motion is viewed from the positive direction of the perpendicular axis.

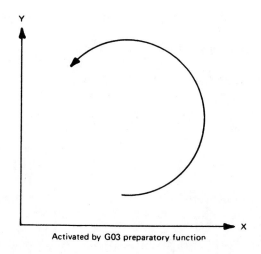

Activated by G03 preparatory function

ARCHITECTURE. The way in which the parts of a system fit and communicate with one another.

ASCII. (American Standard Code for Information Interchange) A data transmission code that has been established as an American standard by the American Standards Association. It is a code in which seven bits are used to represent each character. Formerly USASCII.

AUTOMATION. 1. The implementation of processes by automatic means. 2. The investigation, design, development, and application of methods to render processes automatic, self-moving, or self-controlling.

AUTOSPOT. (Automatic System for Positioning of Tools) An older computer-assigned program for N/C positioning and straight-cut systems, developed in the United States by the IBM Space Guidance Center. At one time, it was maintained and taught by IBM.

AUXILIARY FUNCTION. A programmable function of a machine other than the control of the coordinate movements or cutter.

- Transferring a tool to the select tool position.
- Turning coolant ON or OFF.
- Starting or stopping the spindle.
- Initiating pallet shuttle or movement.

AXIS. A principal direction along which the relative movements of the tool or workpiece occur. There are usually three linear axes, mutually at right angles, designated as X, Y, and Z.

AXIS INHIBIT. A feature of an N/C unit that enables the operator to withhold command information from a machine tool slide.

AXIS INTERCHANGE. The entering of information concerning one axis into the storage location of another axis.

AXIS INVERSION. The reversal of plus and minus values along an axis. This allows the machining of a left-handed part from right-handed programming, or vice versa.

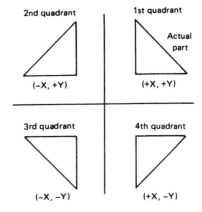

BACKLASH. A relative movement between interacting mechanical parts as a result of looseness.

BATCH PROCESSING. Technique by which items to be processed must be coded and collected into groups prior to processing.

B-AXIS. (or β-axis) The axis of circular motion of a machine tool member or slide about the Y-axis.

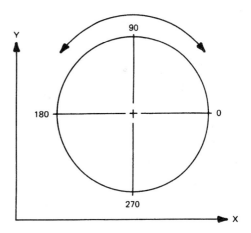

BCD. (binary-coded decimal) A system of number representation in which each decimal digit is represented by a group of binary digits that corresponds to a character.

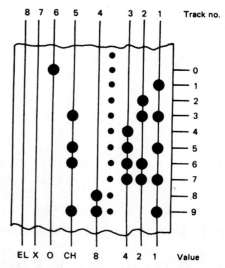

Numbers and letters are expressed by punched holes across the tape for the code or value desired.

BINARY CODE. A code based on binary numbers, which are expressed as either 1 or 0, true or false, on or off.

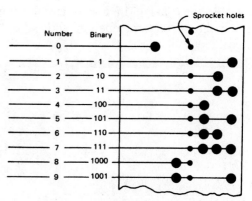

Most computers operate on some form of binary system where a number or letter can be expressed as ON (a hole) or OFF (no hole).

BIT. (binary digit) 1. A binary digit having only two possible states. 2. A single character of a language that uses exactly two distinct kinds of characters. 3. A magnetized spot on any storage device.

BLOCK. A word or group of words considered as a unit. A block is separated from other units by an end-of-block character. On punched tape, a block of data provides sufficient information for an operation.

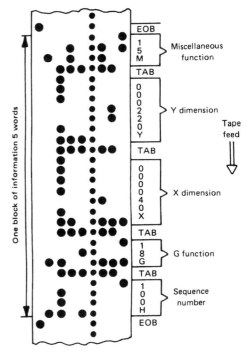

BLOCK DELETE. A function that permits selected blocks of tape to be ignored by the control system, at the operator's discretion, with permission of the programmer.

This feature allows certain blocks of information to be skipped by programming a slash (/) code in front of the block to be skipped. One lot of parts with holes 1, 2, and 3 are required. On another lot, only holes 1 and 3 are required. The same tape could be used for both lots by activating the block delete switch on the second lot and eliminating hole 2. The (/) code would be in front of the block of information for hole 2.

BUFFER STORAGE. A place in which information in a control system or computer is stored for planned use. Information from the buffer storage section of a control system can be transferred almost instantly to active storage (the portion of the control system commanding the operation at that particular time). Buffer storage allows a control system to act immediately on stored information, rather than wait for the information to be read into the machine from the tape reader.

BUG. 1. A mistake or malfunction. 2. An integrated circuit (slang).

BYTE. A sequence of adjacent binary digits usually operated on as a unit and shorter than a computer word.

Eight bits equal one byte. A computer word usually consists of either sixteen or thirty-two bits (two or four bytes).

CAD. (Computer-aided design) The use of computers to assist in phases of design.

CAM. (Computer-aided manufacturing) The use of computers to assist in phases of manufacturing.

CAM-I. (Computer-Aided Manufacturing International) The outgrowth and replacement organization of the APT Long Range Program.

CANCEL. A command that will discontinue any canned cycles or sequence commands.

CANNED CYCLE. A preset sequence of events initiated by a single command. For example, code G84 will perform a tapping cycle by N/C.

CARD-TO-TAPE CONVERTER. A device that converts information directly from punched cards to punched or magnetic tape.

CARTESIAN COORDINATES. A three-dimensional system whereby the position of a point can be defined with reference to a set of axes at right angles to each other.

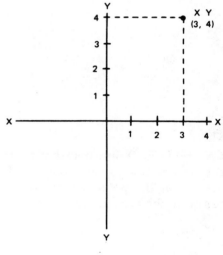

C-AXIS. Normally the axis of circular motion of a machine tool member or slide about the Z-axis. Values along the C-axis are degrees of rotation about the Z-axis.

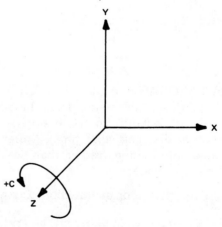

CHAD. Pieces of material removed in card or tape operations.

CHANNELS. Paths parallel to the edge of the tape, along which information may be stored by the presence or absence of holes or magnetized areas. This term is also known as *level* or *track*. The EIA standard 1-in.-wide tape has eight channels.

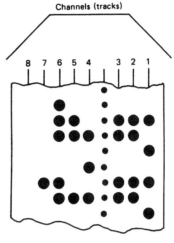

CHARACTERS. A general term for all symbols, such as alphabetic letters, numerals, and punctuation marks. It is also the coded representation of such symbols.

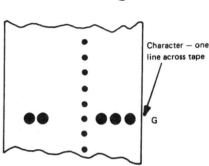

CHIP. A single piece of silicon cut from a slice by scribing and breaking. It can contain one or more circuits but is packaged as a unit.

CIRCULAR INTERPOLATION. 1. The process of generating up to 360 degrees of arc using only one block of information as defined by EIA. 2. A mode of contouring control that uses the information contained in a single block to produce an arc of a circle.

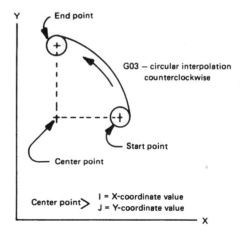

CL DATA. Processor output that contains information regarding cutter location.

CLOSED-LOOP SYSTEM. A system in which the output, or some result of the output, is measured and fed back for comparison with the input. In an N/C system, the output is the position of the table or head; the input is the tape information, which ordinarily differs from the output. This difference is measured and results in a machine movement to reduce and eliminate the variance.

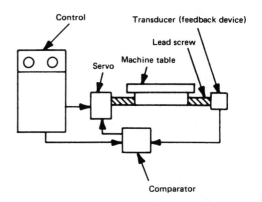

CNC. Computer numerical control.

CODE. A system describing the formation of characters on a tape; it is used to represent information and is written in a language that can be understood and handled by the control system.

COMMAND. A signal or series of signals that initiates one step in the execution of a program.

COMMAND READOUT. A display of the slide position as commanded from the control system.

COMPILE. To generate a machine language program from a computer program written in a high-level source code.

COMPILER. Computer program used to translate high-level source code into machine language programs suitable for execution on a particular computing system.

CONSOLE. That part of a computer used for communication between the computer operator or maintenance engineer and the computer.

CONTINUOUS-PATH OPERATION. An operation in which rate and direction of relative movement of machine members is under continuous numerical control. There is no pause for data reading. (See *contouring control system*)

CONTOURING CONTROL SYSTEM. An N/C system for controlling a machine (for example, during milling or drafting) in a path that results from the coordinated, simultaneous motion of two or more axes.

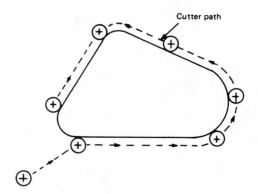

COORDINATE DIMENSIONING. A system of dimensioning based on a common starting point.

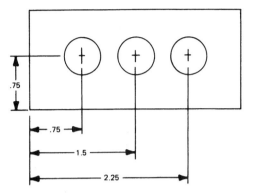

(Also known as absolute dimensioning)

CPU. Central processing unit of a computer. The memory or logic of a computer that includes overall circuits, processing, and execution of instructions.

CRT. (cathode ray tube) A device that represents data (alphanumeric or graphic) by means of a controlled electron beam directed against a fluorescent coating in the tube.

CUTTER DIAMETER COMPENSATION. A system in which the programmed path may be altered to allow for the difference between actual and programmed cutter diameters.

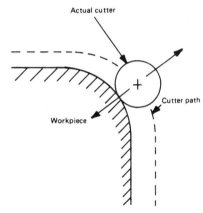

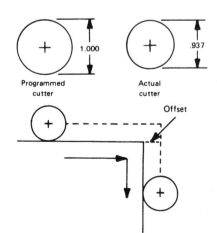

CUTTER OFFSET. The distance from the part surface to the axial center of a cutter (that is, the radius of the cutting tool).

CUTTER PATH. The path defined by the center of a cutter.

CYCLE. 1. A sequence of operations that is regularly repeated. 2. The time it takes for one such sequence to occur.

DATA. A representation of information in the form of words, symbols, numbers, letters, characters, digits, etc.

DATABASE. A comprehensive data file containing information in a format applicable to a user's needs and available when needed.

DEBUG. 1. To detect, locate, and remove mistakes from a program. 2. Troubleshoot.

DECIMAL CODE. A code in which each allowable position has one of ten possible states. (The conventional decimal number system is a decimal code.)

DELETE CHARACTER. A character used primarily to obliterate any erroneous or unwanted characters on punched tape. The delete character consists of perforations in all punching positions.

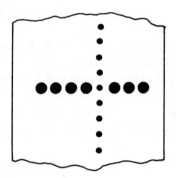

DIAGNOSTIC TEST. The running of a machine program or routine to discover a failure or potential failure of a machine element and to determine its location.

DIGIT. A character in any numbering system.

DIGITAL. 1. Referring to discrete states of a signal (on or off) (a combination of these makes up a specific value). 2. Relating to data in the form of digits.

DISPLAY. A visual representation of data.

DOCUMENTATION. Manuals and other printed materials (tables, magnetic tape, listing, diagrams) that provide information for the use and maintenance of a manufactured product, both hardware and software.

DOWNTIME. Time during which equipment is inoperable because of faults.

DNC. Direct or distributed numerical control.

DWELL TIME. A timed delay of programmed or established duration, not cyclic or sequential. It is not an interlock or hold time.

EDIT. To modify the form of data.

EIA STANDARD CODE. A standard code for positioning, straight-cut, and contouring control systems proposed by the U.S. EIA in their Standard RS-244. Eight-track paper (1-in. wide) has been accepted by the American Standards Association as an American standard for numerical control.

END-OF-BLOCK CHARACTER. 1. A character indicating the end of a block of tape information, used to stop the tape reader after a block has been read. 2. The typewriter function of the carriage return during preparation of machine control tapes.

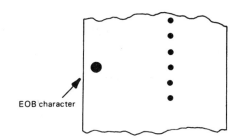

END OF PROGRAM. A miscellaneous function (M02) that indicates the completion of a workpiece. Stops spindle, coolant, and feed after completion of all commands in the block. Used to reset control and/or machine.

END OF TAPE. A miscellaneous function (M30) that stops spindle, coolant, and feed after completion of all commands in the block. Used to reset control and/or machine.

END POINTS. The extremities of a span.

ERROR SIGNAL. Indication of a difference between the output and input signals in a servo system.

EXECUTIVE PROGRAM. A series of programming instructions enabling a dedicated minicomputer to produce a specific output control. For example, it is the executive program in a CNC unit that enables the control to think like a lathe or machining center.

FEED. The programmed or manually established rate of movement of the cutting tool into the workpiece for the required machining operation.

FEEDBACK. The transmission of a signal from a late to an earlier stage in a system. In a closed-loop N/C system, a signal of the machine slide position is fed back and compared with the input signal, which specifies the demanded position. These two signals are compared, and an error signal is generated if a difference exists.

FEED FUNCTION. The relative motion between the tool or instrument and the work due to motion of the programmed axis.

FEED RATE (CODE WORD). A multiple-character code containing the letter F followed by digits. It determines the machine slide rate of feed.

FEED RATE DIVIDER. A feature of some machine control units that allows the programmed feed rate to be divided by a selected amount as provided for in the machine control unit.

FEED RATE MULTIPLIER. A feature of some machine control units that allows the programmed feed rate to be multiplied by a selected amount as provided for in the machine control unit.

FEED RATE OVERRIDE. A variable manual control function directing the control system to reduce the programmed feed rate.

Feed rate override is a percentage function to reduce the programmed feed rate. If the programmed feed rate was 30 inches per minute and the operator wanted 15 inches per minute, the feed rate override dial would be set at 50 percent.

FILE. Organized collection of related data. For example, the entire set of inventory master data records makes up the Inventory Master File.

FIRMWARE. Programs or control instructions that are not changeable (by the user) and that are held in read-only memory (ROM) or another permanent memory device.

FIXED BLOCK FORMAT. A format in which the number and sequence of words and characters appearing in successive blocks is constant.

FIXED CYCLE. See *canned cycle*.

FIXED SEQUENTIAL FORMAT. A means of identifying a word by its location in a block of information. Words must be presented in a specific order, and all possible words preceding the last desired word must be present in the block.

FLOATING ZERO. A characteristic of a machine control unit that permits the zero reference point on an axis to be established readily at any point in the travel.

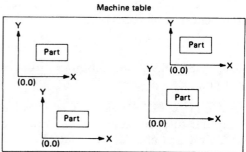

The part or workpiece may be moved to *any* location on the machine table and zero may be established at that point.

FMS. Flexible manufacturing system.

FORMAT (TAPE). The general order in which information appears on the input media, such as the location of holes on a punched tape or the magnetized areas on a magnetic tape.

FULL-RANGE FLOATING ZERO. A characteristic of a numerical machine tool control that permits the zero point on an axis to be shifted readily over a specified range. The control retains information on the location of permanent zero.

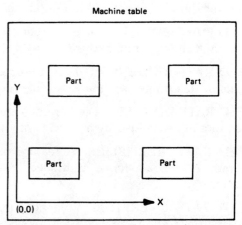

The part or workpiece may be shifted to any position on the machine table, but the actual position of permanent zero remains constant.

GAUGE HEIGHT. A predetermined partial retraction point along the Z-axis to which the cutter retreats from time to time to allow safe XY table travel.

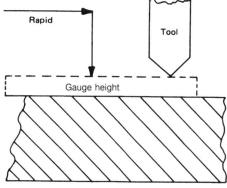

Gauge height, usually .100 to .125, is a set distance established in the control or set by the operator. Gauge height allows the tool, while advancing in rapid transverse, to stop at the established distance (gauge height) and begin feed motion. Without gauge height, the tool would rapid into the part causing tool damage or breakage and potential operator injury.

G CODE. A word addressed by the letter G and followed by a numerical code, defining preparatory functions or cycle types in a numerical control system.

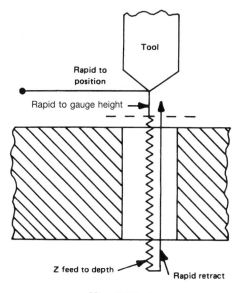

G81 — Drill Cycle

GENERAL PROCESSOR. 1. A computer program for converting geometric input data into the cutter path data required by an N/C machine. 2. A fixed software program designed for a specific logical manipulation of data.

HARD COPY. A readable form of data output on paper.

HARDWARE. The component parts used to build a computer or control system (for example, integrated circuits, diodes, transistors).

HARD-WIRED. Having logic circuits interconnected on a backplane to give a fixed pattern of events.

402 • GLOSSARY

HIGH-SPEED READER. A reading device that can be connected to a computer or control so as to operate on-line without seriously holding up the computer or control.

INCREMENTAL DIMENSIONING. The method of expressing a dimension with respect to the location of the preceding point in a sequence of points.

INCREMENTAL SYSTEM. A control system in which each coordinate or positional dimension, both input and feedback, is given with respect to the previous position rather than from a common datum point, as in the absolute system.

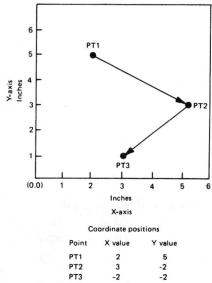

Point	X value	Y value
PT1	2	5
PT2	3	-2
PT3	-2	-2

In an incremental system, all points are expressed relative to the preceding point.

INDEX TABLE CODE. A multiple character code containing the letter b followed by digits. This code determines the position of the rotary index table in degree. (See *B-axis*)

INHIBIT. To prevent an action or acceptance of data by applying an appropriate signal to the appropriate input.

INPUT. Transfer of external information into the control system.

INPUT MEDIA. 1. The form of input, such as punched cards and tape or magnetic tape. 2. The device used to input information.

INTERCHANGEABLE VARIABLE BLOCK FORMAT. A programming arrangement consisting of a combination of the word address and tab sequential formats to provide greater compatibility in programming. Words are interchangeable within the block. The length of a block varies because words may be omitted. (See *block*)

INTERCHANGE STATION. The position where a tool of an automatic tool-changing machine awaits automatic transfer to either the spindle or the appropriate coded drum station.

INTERFACE. Connection or linkage between software modules that are usually in the same mode.

INTERMEDIATE TRANSFER ARM. The mechanical device in automatic tool-changing machine that grips and removes a programmed tool from the coded drum station and places it into the interchange station, where it awaits transfer to the machine spindle. This device then automatically grips and removes the used tool from the interchange station and returns it to the appropriate coded drum station.

INTERPOLATION. 1. The insertion of intermediate information based on an assumed order or computation. 2. A function of a control whereby data points are generated between given coordinate positions.

INTERPOLATOR. A numerical control system device that performs interpolation.

ISO. International Organization for Standardization.

JOG. A control function that momentarily operates a drive to the machine.

LEADING ZEROS. Redundant zeros to the left of a number.

Leading zeros

X + 0062500

LEADING ZERO SUPPRESSION. See *zero suppression*.

LETTER ADDRESS. The means by which information is directed to different parts of the system. All information must be preceded by its proper letter address (for example, X, Y, Z, M).

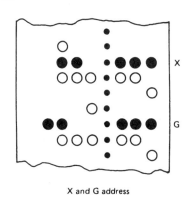

X and G address

An identifying letter inserted in front of each word.

LINEAR INTERPOLATION. A function of a control whereby data points are generated between given coordinate positions to allow simultaneous movement of two or more axes of motion in a linear (straight) path.

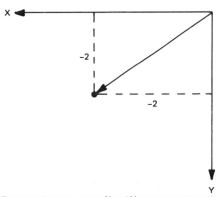

The control system moves X- and Y-axes proportionately to arrive at the destination point.

LOOP TAPE. A short piece of tape, with joined ends, that contains a complete program or operation.

MACHINE CODE. Code obeyed by a computer or microprocessor system with no further translation (normally written in binary code).

MACHINE LANGUAGE. Basic language of a computer. Programs written in a cryptic machine language that only the computer understands.

MACHINING CENTER. Machine tools, usually numerically controlled, capable of automatically drilling, reaming, tapping, milling, and boring multiple faces of a part. They are often equipped with a system for automatically changing cutting tools.

MACRO. A group of instructions that can be stored and recalled as a group to solve a recurring problem.

An APT macro could be as follows:

```
DRILL1 = MACRO/X, Y, Z, Z1, FR, RR
    GOTO/POINT, X, Y, Z, RR
    GODLTA/-Z1, FR
    GODLTA/+Z1, RR
TERMAC
```

X, Y, Z, Z1, FR, and RR would be variables which would have values assigned when the macro is called into action. The variables would be as follows:

    X = X position
    Y = Y position
    Z = Z position (above work surface)
    Z1 = Z feed distance
    FR = feed rate
    RR = rapid rate

The call statement could be:
    CALL/DRILL1, X = 2, Y = 4, Z = .100, Z1 = 1.25, FR = 2, RR = 200

MAGIC-THREE CODING. A feed rate code that uses three digits of data in the F word. The first digit defines the power-of-10 multiplier. It determines the positioning of the floating decimal point. The last two digits are the most significant digits of the desired feed rate.

To program a feed rate of 12 inches per minute in magic-three coding:
1) count the number of decimal places to the left of the decimal. 12 = 2
2) Add magic "3" to the number of counted decimal places. (3 + 2 = 5)
3) write the F word address, the added digit, and the first two digits of the actual feed rate to be programmed. (F512)
4) F512 would be the magic "3" coded feed rate.

This method of feed rate coding is now almost obsolete.

MAGNETIC TAPE. A tape made of plastic and coated with magnetic material. It stores information by selective polarization of portions of the surface.

MANUAL DATA INPUT. A mode or control that enables an operator to insert data into the control system. This data is identical to information that could be inserted by tape.

MANUAL PART PROGRAMMING. The preparation of a manuscript in machine control language and format to define a sequence of commands for use on an N/C machine.

Manual, or hand, programming is programming the actual codes, X and Y positions, functions, etc. as they are punched in the N/C tape.

    H001 G81   X+37500   Y+52500   W01

MANUSCRIPT. A written or printed copy, in symbolic form, containing the same data as that punched on cards or tape or retained in a memory unit.

MEMORY. An organized collection of storage elements (for example, disks, drums, ferrite cores) into which a unit of information consisting of a binary digit can be stored and from which it can later be retrieved.

A computer with a 64,000-word capacity is said to have a memory of 64 K.

MIRROR IMAGE. See *axis inversion*.

MODAL. Pertaining to information that is retained by the system until new information is obtained and replaces it.

MODULE. An interchangeable plug-in item containing components.

N/C. (numerical control) The technique of controlling a machine or process by using command instructions in coded numerical form.

NETWORK ARCHITECTURE. Set of rules, standards, or recommendations through which various computer hardware, operating systems, and applications software function together.

NODE. Point in a network where service is provided or used or where communications channels are interconnected.

NULL. 1. Pertaining to no deflection from a center or end position. 2. Pertaining to a balanced or zero output from a device.

NUMERICAL CONTROL SYSTEM. A system in which programmed numerical values are directly inserted, stored on some form of input medium, and automatically read and decoded to cause a corresponding movement in a machine or process.

OFFSET. A displacement in the axial direction of the tool equal to the difference between the actual tool length and the programmed tool length. (Compare with *cutter offset*.)

ON-LINE. Pertaining to peripheral devices operating under the direct control of the central processing unit.

OPEN-LOOP SYSTEM. A control system that has no means of comparing the output with the input for control purposes (that is, no feedback).

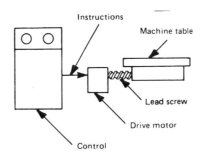

OPERATING SYSTEM. Software that controls the execution of computer programs and that may provide scheduling, input-output control, compilation, data management, debugging, storage assignment, accounting, and other similar functions.

OPTIMIZE. 1. To rearrange the instructions or data in storage so that a minimum number of transfers are required in the running of a program. 2. To obtain maximum accuracy and minimum part production time by manipulation of the program.

OPTIONAL STOP. A miscellaneous function command (M01) similar to Program Stop except that the control ignores the command unless the operator has previously pushed a button to validate the command.

OUTPUT. Data transferred from a computer's internal storage unit to storage or an output device.

OVERSHOOT. The amount by which the motion exceeds the target value. The amount of overshoot depends on the feed rate, the acceleration of the slide unit, or the angular change in direction.

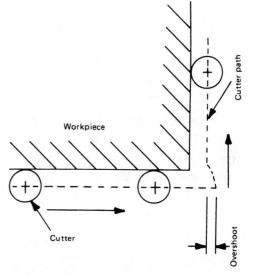

PARABOLA. A plane curve generated by a point moving so that its distance from a fixed second point is equal to its distance from a fixed line.

PARABOLIC INTERPOLATION. Control of cutter path by interpolation between three fixed points, with the assumption that the intermediate points are on a parabola.

PARITY CHECK. 1. A hole punched in one of the tape channels whenever the total number of holes is even, to obtain an odd number (or vice versa, depending on whether the check is even or odd). 2. A check that tests whether the number of ones (or zeros) in any array of binary digits is odd or even

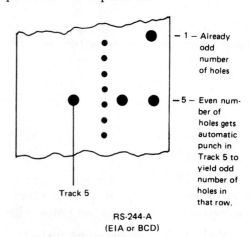

PART PROGRAM. A specific and complete set of data and instructions written in source languages for computer processing or in machine language for manual programming to manufacture a part on an N/C machine.

PART PROGRAMMER. A person who prepares the planned sequence of events for the operation of a numerically controlled machine tool.

PERFORATED TAPE. A tape on which a pattern of holes or cuts is used to represent data.

PLOTTER. A device that draws a plot or trace from coded N/C data input.

POINT-TO-POINT CONTROL SYSTEM. A numerical control system in which controlled motion is required only to reach a given end point, with no path control during the transition from one end point to the next.

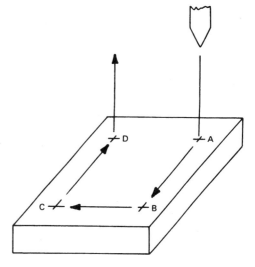

PORT. Point at which data can enter or leave a network, such as the serial or parallel ports in the backs of PCs.

POSITIONING/CONTOURING SYSTEM. A type of numerical control system that has the capability of contouring, without buffer storage, in two axes and positioning in a third axis for such operations as drilling, tapping, and boring.

POSITIONING SYSTEM. See *point-to-point control system*.

POSITION READOUT. A display of absolute slide position as derived from a position feedback device (transducer) normally attached to the lead screw of the machine. See *command readout*.

POSTPROCESSOR. The part of the software that converts the cutter path coordinate data into a form that the machine control can interpret correctly. The cutter path coordinate data is obtained from the general processor based on all other programming instructions and specifications for the particular machine and control.

PREPARATORY FUNCTION. An N/C command on the input tape that changes the mode of operation of the control (generally noted at the beginning of a block by the letter G plus two digits.) (See *G code*)

Some preparatory functions are:

G84 — tap cycle
G01 — linear interpolation
G82 — dwell cycle
G02 — circular interpolation — clockwise
G03 — circular interpolation — counter clockwise

PROCESSOR. A program that translates a source program into object language.

PROGRAM. A sequence of steps to be executed by a control or a computer to perform a given function.

PROGRAMMED DWELL. A delay in program execution for a programmable length of time.

PROGRAMMER. See *part programmer*.

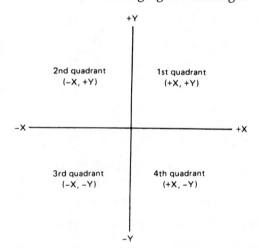

Manual part programming instructions:

```
H001  G81  X+123750  Y+62500   W01
N002       X+105000
N003                 Y+51250          M06
```

Computer part programming instructions:

```
TLRGT, GORGT/HL3, TANTO, C1
       GOFWD/C1, TANTO, HL2
       GOFWD/HL2, PAST, VL2
```

PROGRAM STOP. A miscellaneous function command (M00) to stop the spindle, coolant, and feed after completion of the dimensional move commanded in the block. To continue with the remainder of the program, the operator must push a button.

PROTOCOL. Set of messages with specific formats and rules for exchanging the messages.

QUADRANT. Any of the four parts into which a plane is divided by rectangular coordinate axes in that plane.

RANDOM. Not necessarily arranged in the order of use, but having the ability to select from any location in, and in any order from, the storage system.

RANDOM-ACCESS MEMORY (RAM). Memory device that can be written and read under program control and that is often used as a scratchpad memory by the control logic or by the user for his or her data.

RAPID. To position the cutter and workpiece into close proximity with one another at a high rate of travel speed, usually 400 to 1,200 inches per minute (ipm) or more, before the cut is started.

READER. A pneumatic, photoelectric, or mechanical device used to sense bits of information on punched cards, punched tape, or magnetic tape.

READ-ONLY MEMORY (ROM). Memory device containing information that is fixed and can only be read and not changed.

REGISTER. An internal array of hardware binary circuits for temporary storage of information.

REPEATABILITY. Closeness of, or agreement in, repeated measurements of the same characteristics for the same method and the same conditions.

RESET. To return a register or storage location to zero or to a specified initial condition.

ROUTINE. Set of functionally related instructions that directs the computer to carry out a desired operation. A subdivision of a program.

ROW (TAPE). A path perpendicular to the edge of the tape along which information may be stored by the presence or absence of holes or magnetized areas. A character would be represented by a combination of holes.

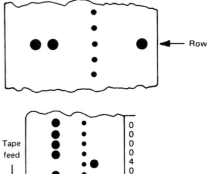

SEQUENCE NUMBER (CODE WORD). A series of numerals programmed on a tape or card and sometimes displayed as a readout; normally used as a data location reference or for card sequencing.

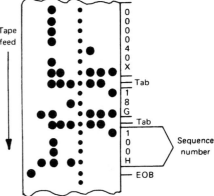

SEQUENCE READOUT. A display of the number of the block of tape being read by the tape reader.

SEQUENTIAL. Arranged in some predetermined logical order.

SERVER. Module or set of modules that performs a well-defined service, such as remote file access or gateway communication, on behalf of another module.

SIGNIFICANT DIGIT. A digit that must be kept to preserve a specific accuracy or precision.

$$X + 00\underbrace{52500}_{\text{Significant digits}}$$

Insignificant digits

SIMULATE. To use one system to represent the functioning of another; that is, to represent a physical system by the execution of a computer program or to represent a biological system by a mathematical model.

SLOW-DOWN SPAN. A span of information having the necessary length to allow the machine to decelerate from the initial feed rate to the maximum allowable cornering feed rate that maintains the specified tolerance.

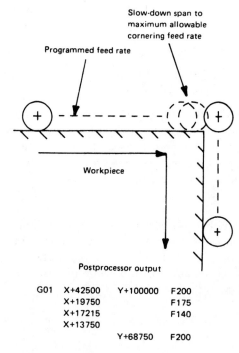

SOFTWARE. Instructional literature and computer programs used to aid in part programming, operating, and maintaining the machining center.

Examples of software programs are:

APT
FORTRAN
COBOL
RPG

SOURCE PROGRAM. Computer program written in a symbolic programming language (for example, an assembly language program, FORTRAN program, or COBOL program). A translator is used to convert the source program into an object program that can be executed on a computer.

SPAN. A certain distance or section of a program designated by two end points for linear interpolation: a beginning point, a center point, and an end point for circular interpolation; and two end points and a diameter point for parabolic interpolation.

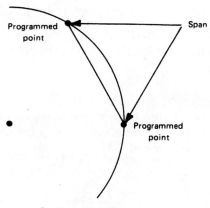

One linear interpolation span

SPINDLE SPEED (CODE WORD). A multiple-character code containing the letter S followed by digits. This code determines the rpm of the cutting spindle of the machine.

STATEMENT. Agreed-upon arrangement of words and/or data accepted by a system to command a particular computer function.

STORAGE. A device into which information can be introduced, held, and then extracted at a later time.

SUBROUTINE. Sequence of callable computer programming statements or instructions that performs frequently required operations (also called a *macro*).

TAB. A nonprinting spacing action on tape preparation equipment. A tab code is used to separate words or groups of characters in the tab sequential format. The spacing action sets typewritten information on a manuscript into tabular form.

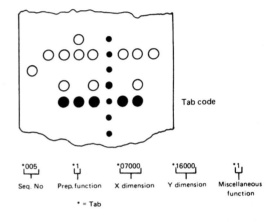

Tab code

TAB SEQUENTIAL FORMAT. Means of identifying a word by the number of tab characters that precede the word in a block. The first character of each word is a tab character. Words must be presented in a specific order, but all characters in a word, except the tab character, may be omitted when the command represented by that word is not desired.

*005 *1 *07000, *16000, *1
Seq. No Prep. function X dimension Y dimension Miscellaneous function

* = Tab

The tab sequential format is, for the most part, obsolete.

TAPE. A magnetic or perforated paper medium for storing information.

TAPE LAGGER. The trailing end portion of a tape.

TAPE LEADER. The front, or lead, portion of a tape.

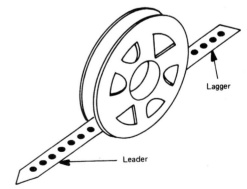

Reel tapes should have a leader and lagger of approximately three feet with just sprocket holes for tape loading and threading purposes.

TERMINAL. Unit in a network, such as a modem or telephone, at which data can be either sent or recieved.

TOOL FUNCTION. A tape command identifying a tool and calling for its selection. The address is normally a T word.

T06 would be a tape command calling for the tool assigned to spindle or pocket 6 to be put in the spindle.

TOOL LENGTH COMPENSATION. A value input manually, by means of selector switches, to eliminate the need for preset tooling; allows the programmer to program all tools as if they are of equal length.

TOOL OFFSET. 1. A correction for tool position parallel to a controlled axis. 2. The ability to reset tool position manually to compensate for tool wear, finish cuts, and tool exchange.

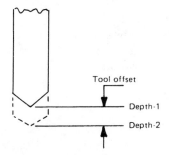

Tool offsets are used as final adjustments to increase or decrease depths due to cutting forces and tool deflection. In this case, a tool offset could be used to increase the drill depth from depth-1 to depth-2.

TOPOLOGY. Physical arrangement and relationship of interconnected nodes and lines in a network.

TRAILING ZERO SUPPRESSION. See *zero suppression.*

TURN KEY SYSTEM. A term applied to an agreement whereby a supplier will install an N/C or computer system such that he or she has total responsibility for building, installing, and testing the system.

USASCII. United States of America Standard Code for Information Interchange. See *ASCII.*

VARIABLE BLOCK FORMAT (TAPE). A format that allows the quantity of words in successive blocks to vary.

Same as word address. Variable block means the length of the blocks can vary depending on what information needs to be conveyed in a given block. See *block.*

VECTOR. A quantity that has magnitude, direction, and sense; it is represented by a directed line segment whose length represents the magnitude and whose orientation in space represents the direction.

VECTOR FEED RATE. The feed rate at which a cutter or tool moves with respect to the work surface. The individual slides may move more slowly or quickly than the programmed rate, but the resultant movement is equal to the programmed rate.

VOLATILE STORAGE. Storage medium in which data connot be retained without continuous power dissipation.

WORD. An ordered set of characters; the normal unit in which information may be stored, transmitted, or operated upon.

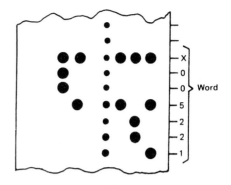

WORD ADDRESS FORMAT. The specific arrangement of addressing each word in a block of information by one or more alphabetical characters that identify the meaning of the word. (See *word*)

WORD LENGTH. The number of bits or characters in a word. (See *word*)

X-AXIS. Axis of motion that is always horizontal and parallel to the work-holding surface.

Y-AXIS. Axis of motion that is perpendicular to both the X- and Z-axes.

Z-AXIS. Axis of motion that is always parallel to the principal spindle of the machine.

ZERO OFFSET. A characteristic of a numerical control machine tool, permitting the zero point on an axis to be shifted readily over a specified range. The control retains information on the location of the permanent zero. (See *full range floating zero* and *floating zero*)

ZERO SHIFT. A characteristic of a numerical machine tool control permitting the zero point on an axis to be shifted readily over a specified range. The control does not retain information on the location of the permanent zero. (See *floating zero*. Consult Chapter 3 for additional details)

ZERO SUPPRESSION. Leading zero suppression: the elimination of insignificant leading zeros to the left of significant digits, usually before printing. Trailing zeros suppression: the elimination of insignificant trailing zeros to the right of significant digits, usually before printing.

Leading zero suppression

X + 0043500

Insignificant digits

Could be written as:

X + 43500

Trailing zero suppression

X + 0043500

Insignificant digits

Could be written as:

X + 00435

INDEX

A-axis, 118, 151
Abrasive waterjet, 317, 317f. *See also* Waterjet machining
Absolute systems *vs.* incremental systems, 123–25
Accountability, group, 61
Accuracy, 255, 256
    CNC machines compared with manual machines, 39–40
    defined, 39
    of parts, 283
Addition, 64–65, 81
Address characters, EIA and AIA national codes for, 379–80
Advanced applications, 315
    composites, 315–17
    part programming technology, 311–15
    in tool performance, 311–13
    waterjet machining, 317–18
Aerospace Industries Association (AIA), 133, 320
    national codes
        address characters, 379–80
        miscellaneous functions, 378–79
        preparatory functions, 377–78
AGVS *See* Automated guided vehicles (AGVS)
AIA. *See* Aerospace Industries Association (AIA)
Air, contaminated, downtime from, 148, 152
Air conditioner, 147
Alertness, 8
Algebra, 72–73
    defined, 72
Algebraic formulas, 85
Alignment positions, 126, 126f
American National Standards Institute (ANSI), 144, 144f, 145, 146f, 152, 265, 278. *See also* American Standard Code for Information Exchange (ASCII)
American Standard Code for Information Exchange (ASCII), 145, 146f, 212. *See also* American National Standards Institute (ANSI)
Angle, 66, 84–85
    acute, 66
    adding or subtracting of, 67
    adjacent, 67
    complementary, 68
    legs of, 67
    obtuse, 66
    opposite, 67
    right, 66
    sides of, 67
    straight, 66, 68
    supplementary, 68
    value of, 68
    vertex of, 67
    vertical, 67
ANSI. *See* American National Standards Institute (ANSI)
Appearance, 48
APT. *See* Automatic programmed tool (APT)
Arithmetic rule with algebraic formula, 72–73
Arm, automated storage and retrieval systems, 364, 364f
ASCII. *See* American Standard Code for Information Exchange (ASCII)
ASRSs. *See* Automated storage and retrieval systems (ASRSs)
Attitude, 52
Automated equipment, safety related to, 8
Automated guided vehicles (AGVS), 361–63
    advantages of, 363
    battery-powered, wire-guided, 363
    types and sizes, 361–63, 362f
Automated storage and retrieval systems (ASRSs), 364f-66
    benefits of, 366
Automatic programmed tool (APT), 320, 321, 327, 374–75
    geometry statement, 323–24
    language, motion statement in, 324
    output from, 321–22
    part programming with, 321
    vocabulary words, 324
Automatic tool changers (ATCs), 222. *See also* Machining centers
Automation, 353–54
    adding of, 354
    conveyors, 367–68, 367f-69f
    defined, 353
    reasons for, 373
    robots, 354–60. *See also* Robot
    single stand-alone machine cell, 353
Axis
    defined, 109
    feed rate, in turning centers, 213
    machine, 106
    machine relationships, 109–13f
    rotary index table of, 240, 242f
    for turning centers, 208, 209–11f
        dimensions, 213

Backing paper, 315, 317
Ballscrews, 106, 106f, 150
B-axis, 119, 151
    for rotary table motion, 240
BCD. *See* Binary-coded decimal system (BCD)
Beacons, 9
Behind the Tape Reader Interface. *See* BTRI (Behind the Tape Reader Interface)
bhn. *See* Brinell hardness number (bhn)
Binary-coded decimal system (BCD), 145, 146f, 152, 212
Bisector, 83
Block, 13, 14f, 156, 165
    full-information, 162
    heel, 294, 294f
    search for, 162
    set-up, 295
    single mode, 204
    support, 294
Block format, flexibility of, 142, 143
Blue cell, 343
Bolts, clamping, 295, 295f, 296–97
Bonding matrix, 315
Boring bar, 265, 266f
    length-to diameter ratio (L/D), 265, 300
Boring cycle, 186f
Boring operations, 264
Boron fibers, 315
Brinell hardness number (bhn), 300–1
BTRI (Behind the Tape Reader Interface), 139
Buffer storage, 102, 102f, 150
    advantage of, 102–3
Business, goal of, 50

CAD (computer-aided design), 324–41, 374
    associativity, 341
    basics of, 332–33
    designing with, 333–36
    programming with, 338–41
    two-dimensional, 333
    visualizing with, 336–37
Calculations, 381
CAM (computer-aided manufacturing), 324, 325–28, 374
    associativity, 341
    basics of, 332–33
    programming with, 338–41
    visualizing with, 336–37
Canned cycle, 182, 189, 227
    programming with, 182–86
Carbide, identification system
    industry standard, 270, 272f–73f

416 • INDEX

Carbide inserts, 201, 279–78, 306
  ceramic, 271
  coated, 271
  grades of, 270, 271
  IC (inscribed circle), 270
  production of, 275
  shape of, 270
  wear, 276
Career opportunities, 32
  methods engineer, 38–39
  operator, 32–34
  operator/programmer, 34–35
  programmer, 35, 36–37
  setup person, 35, 36f
  tooling engineer, 37, 38f
Cartesian Coordinate System, 114, 228. *See also* Rectangular coordinates
Carts, automated, 361–63
Cassette tapes, 135–36, 139
  storage and revision control, 136
Cathode ray tube (CRT) screen, 161, 162f, 314, 314f. *See also* Display screen
C-axis, 119
Cell, 343–45f, 346, 373
  arrangement of, 343, 344
  defined, 343, 344, 373
  levels of, 353
  manufacturing. *See* Manufacturing cell
  multimachine, 346f, 346–47, 347f
  robots used in, 359
Center drills, 257
Centerline cutter location data, 321, 375
Centimeter, 71
Central processing unit (CPU), 313. *See also* Microprocessor
Ceramic (cemented oxide) inserts, 271
Change, Japanese and U.S. view of, 56
Change notice numbers, 199
Channels, 133
Character code recognition, 25
Character coding, 133
Check-out, 201, 202, 248
Chip, 25
  blowing of, 9
  control of, 278, 278f
  removal of, 261, 263
Chip-driver, 260
Chip per tooth, 301
Chipping, 276, 277
Chord of circle, 71
CIM. *See* Computer-integrated manufacturing (CIM)
Cincinnati Hydrotel, 20
Circle, 71
  defined, 71
Circuit boards, vibration of, 148
Circuitry, 26
Circular arc, 234
Circular interpolation, 233–36f, 249
  clockwise, 234, 235f

counterclockwise, 234m 235f
  feed rate for, 233–34
Circumference of circle, 71
Clamp(s)
  H, 293, 295f
  placing of, 284, 286
  plain bar, 291, 292f
  spring-loaded L, 292, 292f
  universal, 293, 293f
Clamping, 9, 283
  procedures for, 294–96
Clamp nuts, tightening of, 296, 296f
CL data, 321, 373
Cleanliness, of toolholders, 279
Clothing, safety related to, 8–9
CNC. *See* Computer numeric control (CNC)
Codes, 13, 157
  adding of, 27
  introduction of, 27
Coding, 145, 152
  learning of, 14
Color displays, 333
Color graphics, 328
Commands, 13
  CAD system, 334
  introduction of, 27
  modal, 180, 189
  nonmodal, 180, 189
Compensation, program for, 181
Competition, 49, 55
Complex machining operations, 41
Composite(s), advanced applications for, 315–17
Composite parts
  forming of, 315
  laying up of, 315
Compressed air, 9
Computer, 101
Computer-aided design. *See* CAD (computer-aided design)
Computer-aided manufacturing. *See* CAM (computer-aided manufacturing)
Computer elements, 26
Computer-integrated manufacturing (CIM), 354
Computer numeric control (CNC), 2, 91, 92
  advanced capabilities, 309–10. *See also* Advanced applications
  fundamentals of, 3–4
  limitations of, 42
  modern, 6f
  student of, 31–32
Computer numeric control (CNC) machines
  compared with manual machines, 39–41, 92
  compared with numeric control machines, 3, 4, 13, 28, 28f
  modern, 21
  operation of, 98
  total number of, 14
Computer numeric control (CNC) machining center, 4f

Computer numeric control (CNC) systems, types of
  beginning from zero, 125–26
  incremental *vs.* absolute systems, 123–25
Computer numeric control (CNC) units
  advanced, 21
  software-based, physical components of, 26
Computing platforms, 325, 326
Constant surface speed (CSS), 169, 170
  facing pass programmed in, 169, 170f
Contouring capability, of machining center, 266
Contour-milling machines, 18
Control, 101
  advancements in, 22, 25, 311
  capabilities, understanding of, 299
  hardware, 107
  hard-wire, 26, 92
  incremental *vs.* absolute systems and, 123
  software-based, 26–29f, 92
Controllers, improvements for, 315
Control unit, 3, 4f, 5f, 6f, 7f, 93
  developments in, 21
  hard-wired, 23, 25
  lab-constructed, 20
  parts of, 25
  types and styles of, 15
Conversational programming, 313. *See also* shop floor programming
Conversion charts, 384, 386
Conveyors
  automated, 367f, 368f–69
  carrousel, 368f, 369
  floor-mounted, 367f, 368f
  overhead-mounted, 367f
  robot arm system, 343f, 346
Coolant, 205, 275, 305
  cutting tool life and, 277, 305
  stopping of, 179, 179f, 180
Cooperation, level of, 46, 47, 47f
Coordinate information, 164. *See also* X words; Y words; Z words
Coordinate measuring machine, 24f, 25
Coordinate system
  machine control in, 116–19
  rectangular, 114–16
Copper, 169
Cost(s), 21
  CNC machines compared with manual machines, 39–41
  of perishable tools, 256
  reduction programs, 56
  related to carbide inserts, 270, 271f
  related to operator/programmer employment, 36
  tooling, 42
  workpiece complexity and, 283
Cost formula, 72
Counterbore, end mill as, 299–300
Counterboring, 185f, 267–68

INDEX • 417

Countersinking, 267, 267f
Co-workers, helping of, 52
CPU. *See* Central processing unit (CPU)
Crane, automated storage and retrieval systems, 366, 366f
Crash, 7, 8
Cratering, 276, 277
CTS. *See* Cutter diameter compensation (CTS)
Customer
    defining of, 49–51, 93
    external, 50–51, 53
    internal, 51, 53
    keeping satisfied, 51–53
Customer-focused, defined, 51, 52, 53
Customer-oriented, defined, 51, 52, 53
Cut
    depth of, 112, 300
    roughing, 301–2
Cutter
    boring, 265
    centerline location data, 321, 373
    centerline of, 165, 165f, 189
    dwelling of, 102
    geometry and position, 237
    life of, increasing of, 301, 302
    milling, sharp corner, 301
    movement of, 116
    oversize and undersize, 237
    radius of, 165
Cutter diameter compensation (CDC), 238–40f, 249
    codes for, 239, 240f
    uses for, 239, 240f
    value, 238, 239
Cutter path
    instructions, 314
    making of, 204
Cutting, 7, 18
Cutting edge, 302
Cutting fluids, 277, 278
Cutting forces, direction of, 302
Cutting inserts, 254
Cutting speed, 8, 169, 170f, 300
    calculation of, 383
    conversion to rpm, 384
    excessive, 301
    selection of, 300–1
Cutting tool, 201, 254
    abnormal performances, 299
    boring and, 265
    circular interpolation, 233–36f
    considerations, 255–56
    countersinking and counterboring, 267–68
    drills, 256–9
    life, 277
    milling and, 266–67
    movements, thru-axis, 20
    radius and length of, 236
    reaming and, 263–65
    rotation, for machining centers, 226

    safety related to, 8
    selection of, 255
    taps, 259–62
    toolholders, 278–79, 280f. *See also* Toolholders
Cycle code, 181
Cycle code number, 182

Damping down, 278
Data, 99–100
    entering the machine control unit, 100
    use of, 98, 101–103
Database, 335
    for geometric translation, 340
Data input and storage, 98, 109, 130
    cassette tapes and floppy disks, 135–37
    DNC (Direct or Distributed Numerical Control), 138–40
    manual, 137, 314, 315
    minimizing amount of, 143–44
    punched tape, 130–35f
Debugging, 157
Decagon, 69
Decimal, addition, subtraction, multiplication and division of, 64–65
Decimal point, 68, 164
    format, 142, 142f, 144
    lining up of, 64, 65
Decimal point programming, 151, 164, 212
    format, for machining centers, 226, 227–29f
    in turning centers, 212–19f
Decimal system, binary coded, 145, 146f, 152
Decision making, 60
Degrees, 66
    measurement of, 68
Dependability, 45
Designer, 326
Design file, 334–35
Design geometry, 332
Design image, graphics creation of, 337
Designing, 195–98
    with CAD system, 333–36
Diameter of circle, 71
Dimension(s), physical, 133
Dimensioning
    absolute, 124, 151
    delta, 123
    incremental, 123, 151
Display screen, 28f, 29f, 112, 113f
Distributed Numerical Control (DNC), 138–40
    advantages of, 139
Division, 64–65, 82–83
Downtime, 145, 147
    causes of, 147–49, 152
Drawing number, 157
Drill(s), 255, 256–59
    carbide insert, 275
    combination, 271, 274f, 274–75

    common twist, 257
    length of, 299, 304
    multiple-diameter, 259
    size, 256, 257
    spade, 257, 258–59
    subland, 259, 259f
    tap, 299
    tolerances, 256, 257f
Drilling cycle, 185f
Drive, servo, electric or hydraulic, 354, 356f
Drive motors, 104–5, 106, 150
    function of, 104
Drowsiness, 11
Dry run, 204, 205, 248

EIA. *See* Electronics Industries Association (EIA)
Electric current, pulses of, 13
Electronic advancements, 22–26
Electronic controls, 2, 3
Electronics Industries Association (EIA), 130, 133, 145, 152, 212
    national codes
        address characters, 379–80
        miscellaneous functions, 378–79
        preparatory functions, 377–78
    RS-274-D, 144, 144f
Employees
    involvement, 61
    tool setup, 201
Employer, needs of, 45–48
End mills, 266–267, 268, 305
    coarse tooth, 301–2
    as counterbore, 299–300
End of Program (M30), 179
Engineer, 326
Engineering, 247
    customer of, 51
Engineering change, 289
Engineering design, 195
Engineering drawing, pencil-and-paper, 333
English, 313
English programming, 28–29f
Enthusiasm, 47–48
Epoxy, 315
Equation, solving of, 74
Equipment, improvement of, 57, 59–60
Errors, 157
    eliminating of, 43, 92
    tape or programming, 64, 321
Executive program, 27, 92
Expensive tools, 43

F, in turning centers, 213
    for machining centers, 226
Facing pass, 169, 170f
Feed, 14, 258
Feedback system, 53, 105, 107–8, 151
    closed-loop, 107, 107f, 108, 109
    open-loop, 107–8

Feed rate, 168–69, 180, 189, 300, 385
  axis, in turning centers, 212
  for circular interpolation, 233–34
  increasing of, 301
Feed rate override switch, 204
Feed value, 381
Feet per minute (fpm), 169
Ferrous metals, 169
Finished goods, 240
Fixture(s), 296. *See also* Jigs
  design and usage, 282–86, 303
  foolproofing of, 286
  hard, 288, 289
  holding, 199–200, 201
  loading of, 36
  number, 157
  placement on machine, 298
Fixture designer, 289
Fixture-mounting surfaces, 295
Fixture supports, 303
Fixturing, 253–4, 255, 282, 283, 305
  considerations, 286–89
  dedicated, 287
  hard, 287
  modular, 287–89, 306
Flame cutters, 21
Flexibility,, 349
Flexible manufacturing cell (FMC), 347
Flexible manufacturing system (FMS), 347–51
  advantages of, 350
Flexowriters, 135
Floppy disks, 135–36, 139
  storage and revision control, 135–6
Flow-line, 198
Flute, reamer, 263, 264f
Flute spacing, 256
FMC. *See* Flexible manufacturing cell (FMC)
FMS. *See* Flexible manufacturing system (FMS)
Foil tape, 133
Foolproofing of fixture, 286
Format tape, 142–43
Formula, 72, 381–85
  solving of, 73
  trigonometry, 74
French curves, 332
Friendliness, 49
Full-floating zero, 125, 126
Functions, 13
  adding of, 26
  changing of, 27
  introduction of, 27
F word, 168, 189

G, 181
  in turning centers, 212
  for machining centers, 225
G02, 233, 234
G03, 233, 234
G40, 239, 240f
G41, 239, 240f
G42, 239, 240f
G98, 186
G99, 186
Gantry machines, 315–17
Gauging systems, 357
G codes, 156. *See also* Preparatory functions, 181, 225
  for CDC, 239, 240f, 241f
  use of, 170
Geometric data, CAD/CAM systems, 340
Geometric representations, CAD, 334
Geometry, 321
  basic, 66–71
  negative rake insert, 300
G functions, 156. *See also* Preparatory functions
Glossary of terms, 386–414
Glove, 10
Good-worker, 47
Gouges, 302
Gouging, 279
Graphical N/C programming, 338
Graphics, 195
  improved, 312
Graphics-based programming, 324–30f, 340, 375
  CAD/CAM, 335–36, 338–41
  interactive systems, 328–30f
Graphics creation of design image, 337
Graphics plot, 204
Graphics terminals, 333
Graphite, 315
Grinding operations, safety related to, 9
G words, 377–78

Hard disk, 312
Hardware, 107, 372
  advancements in, 311–12
  CAD/CAM systems, 332
  flexible manufacturing system, 347
Hard-wired, defined, 26
Hard-wired controls, 27, 28, 92
H block, 165
Heading of manuscript, 157, 188
Heat, 300, 305
  carbide insert wear from, 275, 276, 276f
  downtime from, 147, 152
Heel blocks, 294, 294f
Height adjustment packers, 293, 293f
Holding fixtures, 199–200f, 201
  damage to, 306
  location of parts in, 33, 34
Hole, 131
  blind tapping of, 261
  center drilled, 257, 258f
  cored, 264
  large-diameter, 257
  pattern, 165
  sprocket, 131
  tapping of, 299
Hole producing sequences, 259
Hole punching machines, 21
Horizontal machine, 111, 111f
Housekeeping, 8
H-word address, 162
Hydraulic system, 104, 105
Hydrodynamic machining, 318. *See also* Waterjet machining
Hypotenuse, 70, 75–77

IC (inscribed circle), 270
Identification information, 157
IGES. *See* Initial Graphical Exchange Specification (IGES)
IGES-in capability, 340
IGES-out capability, 340
Illinois Institute of Technology (ITT), 320
Images, storage of, 336
Improvement
  continuous, 55, 62
  eliminating waste, 57–60
  equipment-based, 56–57
  methods of, 55–57
  operational, 56, 57
  philosophy of, 54–55
Inch, conversion
  to metric, 71
  into millimeter, 386
  of millimeter to, 72
Inches per minute (ipm), 168, 381
Inches per revolution (ipr), 168
Incremental systems *vs.* absolute systems, 123–25, 151
Information, communication of, 97. *See also* Communication
Initial Graphical Exchange Specification (IGES), 340
Input media, 18, 130. *See also* Data input and storage, 136, 136f
Inscribed circle. *See* IC (inscribed circle)
Insert nose radius, 300
Inside-diameter (ID) tools, 208, 210f, 211f
Inspection machines, 21
Instructions, 13. *See also* Blocks
  complete block of, 131
  transmitting of, 7
Integrated circuits (ICs), 26
Interactive graphics, 328–30f
Interactive system, 332, 333f
  graphics-based CAD/CAM, 338–41
Interchangeable format, 212
  variable block, in turning centers, 217f–19f
Interpolation
  circular, 233–36f, 249
  defined, 182
  linear, 182, 182f, 234
Interview process, 45f
Inventory, 41, 240
  control, 247
Iron, 169
I word, 234

INDEX • 419

Jacks, support, 294
Japanese
  methods of improvement, 56–58
  view of continuous improvement, 55
Jigs, 282, 283f, 284f, 306. *See also* Fixture
Job
  designing and planning of, 195–98
  self-preparation for, 45–46
Job openings, 45
Just-in-time manufacturing, 198
J word, 234

Kaizen, 55, 56, 93
Kevlar, 315
K word, 234

Language, 15, 155–58f
  processor-based, 320, 322f
Language-based programming, 320–24, 327
Laser machine tool, 24f
Lathe programming, 208–19. *See also*
  Turning centers
Lathe tape input, 144
Laziness, 47
Lead, 169
Leading zero suppression, 166
Lead time, 41
Left-hand helix, 263
Length-to diameter ratio (L/D), 265
  on boring bar, 300
Letter address, 143, 188
Life cycle, 50
Lifting, safety related to, 9
Light beams, photoelectric readers functioning by, 132
Line
  bisector, 83
  end point of, 234
  parallel, 83
  perpendicular, 83
Linear interpolation, 182, 182f, 234
Linear moves, 181–82
Lip height, 256
Loading, 35, 283
Load tape, 27, 93
Logic, 26
  software-based, 27
  software decision, 334
Logic control, 26, 27

M, in turning centers, 213
  for machining centers, 227
M (Mxx), 178
M01 (Optional Stop), 178
Machine
  axes, 106
  axis relationships, 109–13f
  horizontal, 111, 111f
  position of, 109. *See also* Position
  three-axis, 112
  two-axis, 109

  understanding of, 299
  vertical, 111
Machine actions, planning of, 181–82
Machine actuation registers, 101
Machine cell
  single, 344–345f, 346
  stand alone, 353
Machine control unit (MCU), 14, 40, 321
  advancements in, 25
  changes in, 92
  data entering of, 100
  language of, 14, 100
  programming of, 36
Machine control unit (MCU) cabinets, temperature of, 147
Machine table. *See* Table
Machine tape information, 157, 159, 188
Machine time, reduction of, 321
Machine tool
  developments in, 21
  new method of controlling, 2. *See also*
    Numeric control (NC); Computer numerical control (CNC)
  stand-alone, 344, 344f
Machine tool controller, 321
Machine utilization, 41
Machining
  feature-based, 334
  practices, 275
  waterjet, 317–18
Machining centers, 15, 16f, 17f, 168, 255
  contouring capability, 266
  horizontal, 222, 223f, 224, 224f, 240, 249
  innovations, 225
  motions of, 224
  probing used on, 242, 243f
  programming, 225–29
  shops with, counterboring in, 268
  types of, 222–25
  vertical, 24f, 222, 223f, 248
Magnetic tape, 130
Mainframe, 325, 326, 333
Maintenance, 205
Manual data input (MDI), 137, 139, 314
Manual machines
  compared with CNC machines, 40–42, 92
  need for, 42
  safety, 7
Manual tools, 2
Manufacturing
  computer-integrated, 354
  customer of, 51
Manufacturing cell, 343–44
  flexible, 347
Manufacturing engineer. *See* Methods engineer
Manufacturing engineering, 195, 247
Manufacturing run, elements of, 198–202
Manufacturing system, flexible, 347–51
Manuscript, 157, 161, 162f, 188
  creating of, 198
  parts of, 157–59

Manuscript data, 100
Massachusetts Institute of Technology (MIT), 20, 91
Master stop, location of, 11, 10f
Math, 2
  addition, subtraction, multiplication, and division of decimals, 64–65
  basic geometry, 66–71
  introduction to algebra, 72–73
  metric conversion, 71–72
  practicing of, 80–90
  role of, 63–64
  simple right angle triangle trigonometry, 73–77
MCU. *See* Machine control unit (MCU)
MDI. *See* Manual data input (MDI)
Measurement system, selection of, 181
Mechanical readers, 130–131
M30 (End of Program), 179
Menu
  pop-up display, 328
  screen-driven selection, 332, 333f
Metal
  cutting of, 298
  cutting time, 42
  types of, 169
Metal cutting machines, 20
Meter, defined, 71
Methods engineer, responsibilities of, 38–39
Metric conversion, 71–72
Metron, defined, 66
Microchip, 26
Microprocessor, 312f, 313
  complete, 26
  defined, 26
Microprocessor-based controls, 26
Millimeter, 72
  changing to inches, 73
  conversion of inches to, 386
  per minute (mm/min), 168
  per revolution (mm/rev), 168
Milling, 165, 266–67
  CDC value and, 239
  climb, 301
Minutes, 68
Miscellaneous, defined, 178
Miscellaneous functions (M functions), 178–80, 189
  coding
    for machining centers, 227
    in turning centers, 213
  EIA and AIA national codes for, 378–89
Modal commands, 180, 189
Modular circuitry, 26
Modular fixturing, 287–89, 305
Modules, 334
Motion, 104, 106–7, 150, 179
  linear-straight line, 106, 150
Motion statement, in APT language, 324

Motor
    speed of, 104
    stepper, 104, 105
    variable-speed, 105
Mouse, 328, 329f, 340
Movement system, selection of, 181
M00 (Program Stop), 178
M06 (Tool Change), 179
Multimachine cell, 346f, 346–47, 347f
Multiple view capability, of CAD systems, 335
Multiplication, 65, 81
M words, 178, 378–79
Mylar tape, 132

N, in turning centers, 212
    for machining centers, 225
NC. See Numeric control (NC)
Negative rake insert geometry, 300
Nickel, 169, 271
Noncanned cycle, 182
Nonferrous metals, 169
Nonmodal commands, 180, 189
Nonproductive cycle time, 283
Nonservo robots, 354
Non-value-added activities
    defined, 57, 59f, 95
    eliminating of, 57
Numbers, 157
    control of machine tools by, 13. See also Numeric control, 63
Numeric control (N/C), 2. See also Computer numeric control (CNC)
    description of, 13–14
    Direct or Distributed (DNC), 138–139
    fundamentals of, 3–4
    origins of, 20–21
    rectangular coordinates and, 114, 114f. See also Rectangular coordinates
Numeric control (N/C) data. See Data
Numeric control (N/C) machine, 91
    compared with computer numeric control machines, 3, 4, 13, 27, 28f
    modern, 21
    total number of, 14
    vertical, two-axis control on, 116–19
Numeric control (N/C) machine tool
    early model, 20, 22f
Numeric control (N/C) units, hard-wired, 26
N words, 162, 183

O block, 165
Offset, 201, 213. See also Tool offsets
Offset values, 236–38
Operator, 5, 157
    responsibilities of, 33–34, 254
Operator information section of program, 157, 159, 188
Operator/programmer
    responsibilities of, 34–35
    use of math, 63

Optional Stop (M01), 178
Output signal, 13
Outside (OD) diameter tools, 208, 211f, 212
Overtime, 51
O-word address, 162
Oxidation, downtime from, 148–49, 152

Pallet, 344, 345f
Paper tape, 132
Paragraphs, 155, 156
Parsons, John, 21, 311
Part
    accuracy of, 255, 283
    changing of, 199
    commercial, 195, 197f
    cycle of, 344
    defining of, 321
    dimensioning of, 123–25
    features, 315
    geometry, 326, 327, 340
    inventory of, 41
    locating and securing of, 305
    location of, 32, 33
    machining of, 57, 58f
    manufacturing process, 196f
    palletized, in-line system of, 346
    prints, 37, 195, 199
    process plan for, 195, 197f, 198
    programming of
        APT, 321, 323f, 325f
        modern, 313–15
    running of, 205
    setup, 33
    variability of, 40
Part drawing, 125
Parts loader, 33. See also Operator
PCs. See Personal computers (PCs)
Pentagon, 69
Performance capabilities, boosting of, 21
Perishable tools, costs of, 256
Personal computers (PCs), 325, 326, 333
Personal traits, 44
Personnel. See Employees
Photoelectric readers, 130–131, 132
Planner. See Methods engineer
Planning of job, 195–98
Plot, 204, 248
    graphics, 204
    printed, making of, 204
Plus (+) sign, 115, 123
Point-and-click device, 340
Polygon, 69, 69f
    defined, 69
Position, 13
    determination of, 109
Positioning, 25
    error, 124, 125
    in quadrants, 114, 115–16
    X and Y, 109
Postprocessor, 322, 323

Power disturbances, downtime from, 145, 147, 152
Preparatory function, 156, 159, 181, 189
    codes, 212
    EIA and AIA national, 377–78
Preparatory function coding, for machining centers, 225
Primitives, 332
Print reading, 36
Probe, 242
    programming of, 244–45
    stylus, 244, 245
    surface-sensing, 244
Probing, 242–45, 249
Problem, identification of, 53
Process, efficiencies, 53
Process improvement teams, 53
Processing
    complexity of, 283
    improved, 311
Processor
    language-based, 327
    32 bit, 311–12f
Processor-based languages, 320, 322f
Process plan, 195, 197f, 198, 199, 247
    changing of, 199
Product, life cycle of, 51
Production planning, 41
Production run, 204–5
Production tools, jigs and fixtures, 282–86. See also Fixture; Jigs
Program, 5, 13, 156
    changing of, 199
    creating of, 198
    defined, 13, 14, 91
    dry run, 204
    example of, 14f
    executive, 27
    manuscript, 161, 162f, 188
    moves making up, 5
    quantities generated, 224
    role of setup person in, 35
    running of, 248
    in single block mode, 204
    starting of, 198
    storage of, 14
    internal memory for, 139
Program manuscript, 32, 157
Programmed word, sign of, 164
Programmer, 5, 157, 159. See also Operator/programmer
    immediate feedback to, 339
    part, questions asked by, 100
    responsibilities of, 35, 36–37
    wages of, 35
Programming, 2, 99–100, 125, 157
    absolute, 124, 125, 151
    baseline, 124
    with CAD/CAM systems, 338–41
    concept of, 14–17f
    English, 27–29f

flow-line, 198
graphics-based, 324–29f
incremental, 123
language-based, 320–24, 327
at the machine, 35, 36
by methods engineer, 38, 39f
N/C machines compared with CNC machines, 14–15
negative, 126
off-line, 34, 35, 157
positive, 126
of probe, 244–45
safety related to, 7–10f, 11
visualization and, 17
Programming errors, 321
Programming functions, 233
circular interpolation, 233–36f
cutter diameter compensation, 238–40f
probing, 242–45
rotary index tables, 240, 241–42f
tool offsets, 26–37, 238f
Program Stop (M00), 178
Prove-out, 202, 205, 248
inspecting the first part, 203–6
methods for, 203, 204–5
purpose of, 203
rules to follow, 205–6
time spent on, 202
Punched cards, 130
Punched tape, 130–35f
materials, 132
size of, 133, 133f
storage and revision control, 135–136
Punctuation, 156
Pythagorean theorem, 73–77

Quadrants, 114, 114f
positioning in, 114, 115–16
Quadrilateral, 69
Quality, 54
improvement of, 54
programs for, 55

Radius of circle, 71
Rapid-to-position move, 172
Ratio, 74
trigonometric, 74
Raw material, 240
Readers, 130–135f
Reading cycle, stopping of, 178
Readout screen, 3, 5
Reaming, 263–65, 264f
Rectangular coordinates, 114, 115f, 151
Register
actuation, 109, 150
machine actuation, 101, 103, 109, 112
temporary holding, 102
X and Y, 166
Reliability, 46
CNC machines compared with manual machines, 40

Repeatability, CNC machines compared with manual machines, 40
Resolution, 106, 107, 151
Revision, 199
defined, 247
Revision numbers, 199, 247–48
Revolutions per minute (rpm), 168, 169, 189
cutting speeds conversion to, 384
determining of, 381
Right-hand helix, 263
Rigidity, 291, 305
Robot, 346
application of, 354–60
axes of movement, 354, 357f
benefits of using, 357–58
conveyor arm system, 346
mounting of, 357
types of, 354–57
Rotary axes of motion, 118, 151
Rotary index table, 224, 240, 241–42f, 249, 288
Rotation
negative, 66
positive, 66
RS-232 Communications port, 139
Run. See Manufacturing run
R word, 172, 183

S, in turning centers, 213
for machining centers, 226
Safety, 5–11, 91, 291, 305
manual machine tools compared with N/C and CNC machine tools, 7, 11
rules, 8–9, 11
Safety glasses, 8
Safety guards, 9, 11
Safety (steel toe) shoes, 8
Satisfaction, customer, 51–53
Schematic, 183
Schools, vocational and technical, 32
Scrap, reduced, 41
Screen-driven menu selection, 332, 333f
Seconds, 68
Self-preparation, 45–46
Sentences, 155
Sequence number, 161–2
coding, for machining centers, 225
for turning centers, 212
Servo, 104–5, 150, 313
alternating current (AC), 104, 105, 150
direct-current (DC), 104, 105, 150
hydraulic, 104, 105, 150
Servo-driven robots, 354, 357–59
Setup(s), fundamentals of, 291–94
Setup instructions, 198–99, 201
Setup person, responsibilities of, 35, 36f
SFP. See Shop floor programming (SFP)
Shaft-cell, 343
Shapes
CAD system, 334
complex, 40

Shearing action, 276
Shell reamer, 263, 264f, 265
Shock-absorbing pads, 148
Shoes, 8
Shop floor programming (SFP), 313–15
Single block mode, running program in, 204
Skills, acquiring of, 44, 92
Slant-bed design, 208, 210f
Software
CAD/CAM systems, 332, 333
flexible manufacturing system, 347, 349
Software-based control, 26–27, 28f, 92
Software decision logic, 334
Solids, 332
Space requirements, 41
Spade drill, 257, 258–59
Special functions, activation of, 178
Speed
determination of, 107
Speedometer, 107
Spindle, 9, 278
direction of, 111
movement of, 172
speeds, 169–71, 189
in turning centers, 213
vertical, 15f, 18
Spindle profiler, 5f
Spotting drills, 257
Sprockets, 131
Stacker crane, 364, 364f
Standardization, 133
Steel, 169, 258
Stepper motors, 104, 105
Stick figures, 337
Stock market analysts, 57
Stock removal, 314
Storage media, 135, 136f
Strength, related to carbide inserts, 270, 271f
Studying CNC, 31–32
Stylus, 244
Subland drills, 259, 259f
Subtraction, 64–65, 81
Suggestions, making of, 52, 53
Suppress, defined, 166
Surface model, 336, 336f, 337
Surface-sensing probe. See Probe
S-word address, 169, 189
Symbol, 323

T, in turning centers, 213
for machining centers, 225
Table, 294
cleaning of, 199, 201
movement of, 117
positioning of part and fixture in relation to, 286–87, 289
travel speeds, 106, 151
Tap(s), 259–63
cam-point, 260
drills, 299
fluteless, 262f, 263

Tap(s) (cont.)
  hand, 260, 260f
  life of, reduction, 263
  machine screw, 260
  spiral fluted, 261, 262f
  spiral pointed, 260–61, 261f
Tape
  cassette, 135–36
  coding systems, 145
  errors, 321
  format, 151
  formats, 142–44
  magnetic, 130
  punched, 130–35f
Tape-punching equipment, computer, 133, 135f
Tape readers, 130–35f
  classification, 130–31
Tape search feature, 161, 162, 188
Tapping, 299, 304
  feed rates, 385
Tapping lubricants, 263
T-bolts, 295, 295f. 296
Team, 60–61, 93
  defined, 60
  formation of, 61
  function of, 61
  participation on, 60–61
"Team player," 46
Threads per inch-to-lead, 385
Three-dimensional (3-D) models, 336, 337
Time-in-the-cut, 18, 256
Tin, 169
Titanium carbide, 276
Titanium-nitrate (TIN) coating, for carbide inserts, 271
Tool
  adjustment of, 36
  automatic programmed, 320, 321, 373–74
  care for, 299
  changes, 214
  checking of, 298
  design, 198
  design flow line in, 196f, 198
  lengths and diameters, 181, 287
  life, 256
  loading of, 36
  manual removal of, 10
  motion of, 321
  performance, advancements in, 311–13
  perishable, cost of, 256
  selection of, 298–302
  standard, 300
  usage, 304
  warehouse, ASRSs for, 366
  wear, 237
Tool Change (M06), 180
Tool changers, automatic, 322. See also Machining centers
Tool designer, 289

Tool drum, 201
  storage space, 267
Toolholder, 210, 278–79, 280, 306
  floating, 264
  lathe, 303
  spindle, 303
  tapered V flange, 278
  turning center, 279, 280f
Toolholder shank, 301
Tooling, 253–4
  by methods engineer, 39, 39f
  practices, 299–303, 307
  principles, 16
  setup, 298
Tooling engineer, responsibilities of, 38, 39f
Tooling offsets, 201
Tool matrix pockets, 278
Tool number coding, for machining centers, 226
Tool offset, 249
Tool offsets, 210, 236–37, 238f
  rough and finish passes, 237
  uses of, 237, 238f
Tool path, 204
  calculated, 340
Tool path editing, 329
  interactive graphical, 340
Tool-setting pads, 255
Torsional rigidity, 257
Touch trigger, 242, 242f. See also Probe
Tracks, 133
Training, acquiring of, 45, 93
Trammed in, 126
Transistor, 23
Trial-and error, 289, 290
Triangle, 69, 70
  acute, 70, 70f
  classification, 69, 70f
  equilateral, 70, 70f
  general, 69, 71f
  isosceles, 69, 70f
  obtuse, 70, 70f
  right, 70, 70f
    trigonometry, 73–77
  scalene, 69
Trigonometry, 382
  formulas and triangles, 75–77f
  problems and solutions, 75, 77f
  simple right triangle, 73–77
Tube benders, 21
Tungsten carbide, 276
Turning centers, 168, 248
  axes, 209, 212f
  horizontal, 15, 208, 210f, 211f
  OD and ID operations, 210
  probing used on, 242, 243f
  programming of, 212–19
  slant-bed design, 208
  toolholders, 279, 280f

  types of, 208–12f
  vertical, 15, 15f, 16f, 208, 209f
  words used on, 212–13
Turret station, in turning centers, 213–14
Two-axis control, on vertical numeric control machine, 116–19
Two-axis machine, 109

Unclamping, 283
U.S.
  Air Material Command, 20
  methods of improvement, 56–58
Unloader, 33. See also Operator
Unloading, 283
Urgency, sense of, 46, 47, 48

Value, 50, 51
Value-added activities
  defined, 58, 59f, 94
Vehicles, automated guided. See Automated guided vehicles (AGVS)
Vertical machines, 111
  two-axis control on, 116–19
Vibration, downtime from, 148, 152
Vises, 295, 296, 298
Vision systems, 358
Visualization, 15
  with CAD/CAM, 336–37
  in three-dimensions, 38

Wall Street, 57
Warning lights, 9
Waste
  causes of, 60–61
  elements of, 58, 59
  eliminating of, 58–61
Waterjet machining, 317–18
Web, centrality, 256
Wire-frames, 337, 337f
Word(s), 13, 155, 159
  in APT geometry statement, 323
  in blocks, 13, 14f
  learning of, 14
  letter address for, 143
  for machining centers, 225–29
  programmed, 164
  sequence number, 161, 162f
  used on turning center, 212–13
Word address, 144, 144f, 162, 212
Word address format, 142, 143, 143f, 144
Workholding devices, 42
Work-in-progress, 240
Workpiece, 214
  clamping of, 284, 287
  complexity of, 283
Workpiece material, knowledge of, 301
Work platforms, 10
Workstations, 325, 326, 333
Work table
  on horizontal machine centers, 240

X

negative, 248
positive, 248
in turning centers, 213
   for machining centers, 225
value of, 113
X-axis, 109, 110f, 111, 119, 151
   for machining centers, 226
   movement, 109, 110f, 111
   negative, 209
   positioning of, 137
   positive, 209
   rectangular coordinates and, 113, 115, 116
   for turning centers, 209
   on vertical N/C machine, 116–19
X coordinate information coding
   for machining centers, 225
X coordinate value
   in circular interpolation, 234
X direction
   in machine centers, 228
   negative, 113, 115
X motion, stopping of, 179
X move
   negative, 123
   positive, 123
X value, positive, 118
X word, 188, 189, 234
   programming with, 164–65

Y

for machining centers, 225
value of, 113, 115
Y-axis, 109, 110f, 111, 151
   for machining centers, 226
   movement, 109, 110f, 111
   positioning of, 137
   on vertical N/C machine, 116–19
Y coordinate value, 234
Y direction
   in machine centers, 228
   negative, 116
Y motion, stopping of, 179
Y move
   negative, 123
   positive, 123
Y value, positive, 118
Y word, 188, 189, 234
   programming with, 164–65

Z

negative, 248
positive, 248
in turning centers, 213
   for machining centers, 225
Z-axis, 111, 112, 113, 119, 151, 172
   coordinate information coding, 225
   depth of, 112
   horizontal, 119
   movement, on vertical N/C machine, 117–19
   negative, 209
   positioning of, 137
   positive, 209
   for turning centers, 209
Z coordinate input, 172, 189
Z coordinate value, 234
Z depth, 186
   programmable, 172–74
Zero, insignificant, 166
Zero point, 113
Zero position, beginning from, 125–26
Zero shifting, 126
Zero suppression, 165–66
Zinc, 169
Z motion, 172, 174
   179, 179
Z move, 118
   negative, 123, 172, 189
   positive, 123, 172, 189
Zooming features, of CAD systems, 335
Z-value, 173
Z word, 164, 172, 188, 189, 234